Green Energy and Technology

Climate change, environmental impact and the limited natural resources urge scientific research and novel technical solutions. The monograph series Green Energy and Technology serves as a publishing platform for scientific and technological approaches to "green"—i.e. environmentally friendly and sustainable—technologies. While a focus lies on energy and power supply, it also covers "green" solutions in industrial engineering and engineering design. Green Energy and Technology addresses researchers, advanced students, technical consultants as well as decision makers in industries and politics. Hence, the level of presentation spans from instructional to highly technical.

Indexed in Scopus.

Indexed in Ei Compendex.

Yelizaveta Chernysh · Viktoriia Chubur ·
Hynek Roubík

Sustainable Biotechnology and Environmental Protection

A Four-pronged Strategy

Springer

Yelizaveta Chernysh
Department of Sustainable Technologies, BioResources and Technology Division
Czech University of Life Sciences Prague
Prague, Czech Republic

Department of Ecology and Environmental Protection Technologies
Sumy State University
Sumy, Ukraine

Viktoriia Chubur
Department of Sustainable Technologies, BioResources and Technology Division
Czech University of Life Sciences Prague
Prague, Czech Republic

Hynek Roubík
Department of Sustainable Technologies, BioResources and Technology Division
Czech University of Life Sciences Prague
Prague, Czech Republic

ISSN 1865-3529 ISSN 1865-3537 (electronic)
Green Energy and Technology
ISBN 978-3-031-89435-0 ISBN 978-3-031-89436-7 (eBook)
https://doi.org/10.1007/978-3-031-89436-7

This Springer imprint is published by the registered company Springer Nature Switzerland AG
The registered company address is: Gewerbestrasse 11, 6330 Cham, Switzerland

Preface

As the world faces unprecedented challenges posed by environmental degradation, it becomes increasingly crucial to seek solutions that not only mitigate harm but also restore and protect the ecosystems upon which all life depends. Sustainable biotechnology offers a unique and transformative approach to addressing these challenges, providing powerful tools to repair and improve the environment while supporting continued human progress.

In this book, *Sustainable Biotechnology and Environmental Protection: Four-Pronged Strategy*, we explore the crucial intersection between biotechnology and environmental protection. Our aim is to offer an in-depth understanding of how biotechnological processes can be harnessed for environmental benefit, offering novel ways to tackle pollution, restore damaged ecosystems, and create sustainable futures for both urban and rural areas.

The four-pronged strategy presented in this book—minimizing *waste generation, reducing toxicity, improving disposal methods, and utilizing waste as a resource*—reflects the complexity of our environmental challenges and the depth of innovation required to overcome them. Each chapter draws on research and real-world applications to demonstrate how sustainable biotechnology can be applied to specific environmental issues. From waste treatment and pollution management to the restoration of biodiversity and the mitigation of climate change, there is clear potential of biotechnology to serve as a cornerstone of a sustainable future.

I have had the privilege of working alongside researchers, scientists, and policymakers who are dedicated to improving the environment through innovative and sustainable methods. And this book, which we have prepared for you with **Yelizaveta Chernysh** and **Viktoriia Chubur**, is built upon experiences and knowledge of the field, which we have been gathering over the years. This book serves as both a comprehensive reference and a call to action for those in the field of environmental biotechnology, as well as anyone invested in creating a more sustainable world. I am proud to contribute to this growing body of knowledge with my colleagues and to share the insights and findings that will help pave the way for a cleaner, healthier, and more sustainable planet.

As we continue to face the global challenges of the 21st century, I encourage you, the readers, to explore the solutions presented within these pages and consider how you can contribute to the ongoing efforts of protecting and nurturing our environment for generations to come.

Prague, Czech Republic Hynek Roubík

Competing Interests The authors have no competing interests to declare that are relevant to the content of this manuscript.

Introduction

The industrialization of economic activities has led to numerous adverse effects on the environment, ranging from the degradation of soil and natural habitats to the pollution of air and water. It has been a driving force behind increased energy consumption, the deterioration of agricultural land, the loss of biodiversity in urban areas, and the generation of new stationary and mobile sources of air emissions. Moreover, the discharge of pollutants into water bodies, deforestation, and the loss of forests have further exacerbated the situation. These environmental consequences contribute to a range of issues, such as altered water regimes, changes in landscapes, loss of numerous species of fauna and flora, and a negative impact on gas exchange in the atmosphere. For example, pollution from sulphur dioxide leads to acid rain, while nuclear energy poses a risk of radioactive contamination in the event of accidents. The construction of hydroelectric power plants often results in the flooding of agricultural lands, a reduction in fish resources, and a deterioration in the self-purification capacity of water bodies. In the world, solid and liquid waste represent a significant environmental challenge, with each resident contributing from 0.11 to 4.54 kg of various waste daily, much of it organic or synthetic.

Environmental pollution in many regions has reached critical levels, making it imperative to find sustainable solutions. This book highlights the role of **Sustainable Biotechnology and Environmental Protection** in providing solutions for the protection and restoration of the environment. Environmental biotechnology is increasingly recognized as a vital tool for addressing ecological challenges and promoting sustainability. By focusing on processes that mitigate environmental harm, this book aims to shed light on the importance of sustainable biotechnology in achieving a safe, resilient, and sustainable environment.

The central focus of this book is on the integration of **biotechnological solutions** in environmental protection, emphasizing a **four-pronged strategy**:

- **Minimizing the amount of waste generated.**
- **Reducing the toxicity of the wastes.**
- **Finding more satisfactory methods of waste disposal.**
- **Reconsidering waste as a resource for valuable raw materials.**

Through these strategies, biotechnology presents innovative ways to mitigate environmental harm and even turn waste into a resource, thereby contributing to a circular economy.

The industrial activities responsible for environmental degradation are intricately linked to the pressing need for ecological restoration. Sustainable biotechnology offers a variety of solutions, not only in waste management but also in the recovery of resources and the reduction of harmful emissions. This book provides an in-depth look into the biological systems, technologies, and strategies that underlie the field of environmental biotechnology. By examining key biological agents, such as microorganisms, microbial consortia, and enzymes, we delve into the processes of bioremediation and resource recovery.

The book also addresses the engineering implementation of these biotechnologies, focusing on the infrastructure and systems needed for large-scale application in various sectors. Additionally, the concept of **bioeconomy** is explored, examining how biotechnology can be integrated into economic models that value natural resources while minimizing waste and ecological impact. The integration of biotechnologies into urban development offers significant potential for solving environmental problems, implementing preventive measures, and reversing the harmful effects of pollution.

The interdisciplinary nature of environmental biotechnology brings together principles from microbiology, chemistry, ecology, and engineering, creating opportunities for sustainable solutions to global environmental challenges. Achievements in ecological science have directly influenced the development of modern biotechnology and continue to drive progress in the field of environmental protection.

This book is structured around a four-pronged strategy to promote environmental sustainability, which encompasses both the scientific and practical applications of biotechnology in addressing environmental issues. By exploring these themes and highlighting cutting-edge research, we aim to provide a comprehensive guide for understanding and advancing the role of biotechnology in environmental protection.

Ultimately, the goal of this book is to provide an insightful exploration of how sustainable biotechnology can serve as a cornerstone for cleaner, more resilient ecosystems, contributing to the health and well-being of both natural environments and human communities alike.

Contents

About the Authors

Assoc. Prof. Dr. Yelizaveta Chernysh is from BioResources and Technology Division, Department of Sustainable Technologies, Czech University of Life Sciences Prague. She works at Department of Ecology and Environmental Protection Technologies of Sumy State University. Assoc. Prof. Dr. Yelizaveta is Graduate of the Sumy State University (SumDU, Ukraine), where she obtained a Candidate of Science and Doctor of Science in ecological safety and a master's in ecology and environmental protection. She also holds the position of Associate Professor at the Department of Ecology and Environmental Protection Technologies of Sumy State University. She is focused on interdisciplinary research in waste recycling and bioenergy. In particular, her research interests include using phosphogypsum as a sustainable feedstock for biotechnological production, nonlinear processes in ecology, and wastewater management. e-mail: chernysh@ftz.czu.cz

Ing. Viktoriia Chubur Ph.D. is from BioResources and Technology Division, Department of Sustainable Technologies, Czech University of Life Sciences Prague. Viktoriia is Graduate of SumDU, where she obtained a Ph.D. in Environmental Protection Technologies and a master's and bachelor's in Ecology and Environmental Protection. Her research focuses on the environmentally safe utilization of waste for energy purposes in environmental protection technologies, with a specific focus on the anaerobic digestion of waste for biogas production. She is also engaged in research on phosphogypsum processing. e-mail: chuburv@ftz.czu.cz

Assoc. Prof. Ing. Hynek Roubík Ph.D. is BioResources and Technology Division Director and Dean at the Faculty of Tropical AgriSciences at the Czech University of Life Sciences Prague. He is Founder of the BioResources and Technology (BRT) Division (Czech University of Life Sciences Prague) and Director. He deals not only with aspects related to biogas but also with waste management issues in general. Assoc. Prof. Hynek Roubík participated in numerous research and development projects throughout the world as the project leader or expert (i.e. Czechia, Vietnam, Cambodia, Ukraine, Uzbekistan, Georgia, Sri Lanka, Indonesia and others). He does research, especially in Waste Management, Environmental and Ecological

Engineering, and is Editorial Board Member of several journals. He is Author of over 150 peer-reviewed (indexed) research papers. He is also Editor or Associate Editor for several scientific journals. He worked as Expert on demand for several organizations, such as the UNDP and the European Commission. He is Member of the Research, Development and Innovation Council (R&D&I Council) of Czechia, a professional consultancy body of the Government of Czechia in research, experimental development and innovation. In addition, he is Second Youngest Member of the Czech Academy of Agricultural Sciences and Bioeconomy platform of Czechia and was appointed as one of the youngest associate professors in Czechia. e-mail: roubik@ftz.czu.cz

Chapter 1
Introduction to Environmental Biotechnology

Abstract The development of biotechnology is important for the establishment of sustainable development goals. An important direction of implementation of bioprocesses in environmental protection is realized in environmental biotechnologies. In particular, they relate to the introduction of technologies of waste bioprocessing, restoration of fertility of polluted soils, purification of water bodies, and reduction of greenhouse gas emissions.

1.1 Subject, Goals and Objectives of Environmental Biotechnology

Biotechnology is critical to achieving most of the Sustainable Development Goals (SDGs), largely because of the central role of microorganisms in providing and regulating ecosystem services. For example, as indicated in a study by Timmis et al. [1] microbial communities play a vital role in primary productivity, nutrient cycling, and disease regulation, thereby affecting human, animal, and plant health and thus farm productivity and food security.

Microorganisms and microbial technologies are now increasingly used for disease prevention and treatment as well as to sustainably improve farm profitability, productivity, and nutrient quality, directly contributing to SDGs 1, 2, and 3. Microorganisms also play an important role in climate regulation by contributing to the production and consumption of greenhouse gas (GHG) emissions, as well as creating an environment (e.g. soil structure) for the growth of other organisms on land and in water. A number of microbial technologies are being used or are under development to remove greenhouse gases and treat wastewater for consumption. Microorganisms are the source of an exceptional range of chemicals and chemical catalysts, bioactive substances, biomaterials, and important forms of bioenergy, including hydrocarbons and electricity. They are key agents in the removal and recycling of pollutants. Microorganisms have been a source of industrial productivity throughout human history and are now major drivers of the bioeconomy and industry, worth several trillions of dollars, contributing directly to SDGs 8 and 9.

Y. Chernysh et al., *Sustainable Biotechnology and Environmental Protection*, Green Energy and Technology, https://doi.org/10.1007/978-3-031-89436-7_1

Modern biotechnologies play a crucial role in conserving biodiversity, with plant tissue culture (micropropagation) serving as a key tool in these efforts. As we strive to meet the SDGs, it's essential to consider the potential of biotechnology to harness the power of microorganisms and plant tissue culture in preserving biodiversity while also conserving ecosystem services and promoting sustainable agriculture practices that support human well-being [2]. Figure 1.1 illustrates the role of microbial technologies in advancing the Sustainable Development Goals.

Thus, a characteristic feature of biotechnology is the variety of applications it can address and the wide range of human activities and needs for which it is and can be applied. Because the Sustainable Development Goals have very diverse and complex

Fig. 1.1 Microbial biotechnology pathways and their links to the sustainable development goals

components and requirements, biotechnology can significantly contribute to many levels and in various areas to achieve sustainable development.

One of the most important areas of biotechnology is to limit the scale of pollution on our planet caused by industrial, agricultural and household waste and toxic components of car exhaust. This is how a new scientific discipline emerges—***environmental biotechnology ("grey" biotechnology)***, which is the latest approach to the protection and preservation of the environment by sharing the achievements of biochemistry, microbiology, genetic engineering, ecology and chemical technologies [3].

Thus, the specific application of biotechnological methods to solve environmental problems, such as waste recycling, water purification, and pollution prevention, is the ***subject of environmental biotechnology***. Figure 1.2 shows the main directions of development of environmental biotechnology, which are directly the subject of its study.

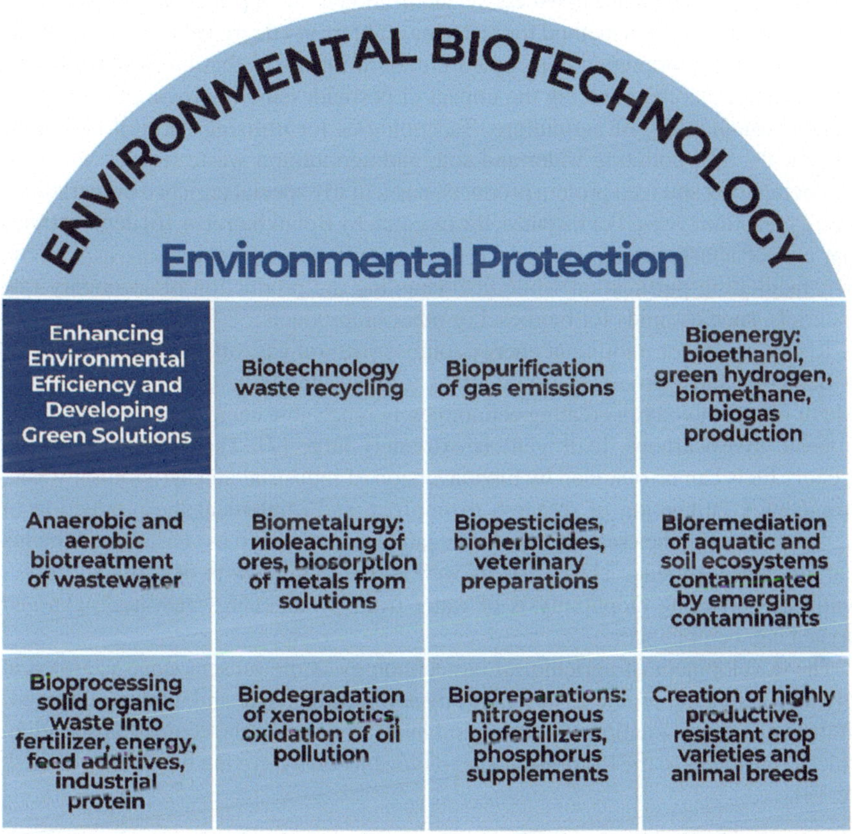

Fig. 1.2 Classification of directions of development of environmental biotechnology

The range of problems solved by environmental biotechnology is extremely wide—from developing and improving methodology for comprehensive chemical and biological research of ecosystems near the sources of technogenic impact to developing technologies and recommendations for soil reclamation, biological water and air purification and biosynthesis. Within the framework of this scientific direction, technologies of soil reclamation, biological purification of water and air and biosynthesis of drugs that compensate for the harmful effects of the changed environment on humans and animals are being developed.

Modern scientific research aims to create waste-free technologies, obtain easily destructive polymers, particularly of biogenic origin, and search for new active microorganisms—polymer destroyers (polyethylene, polypropylene, polyvinyl chloride). The study by Gilani et al. [4] underscores the potential of microorganisms in addressing pollution, focusing on optimizing CO_2 and wastewater concentrations for cultivating cyanobacteria, which proved effective for both bioremediation and biodiesel production. Biotechnology efforts are aimed at combating pesticide pollution—a consequence of excessive and irrational use of pesticides. According to Arora and Fatima [5], a method for restoring soil fertility degraded by pesticides was developed using biopreparations based on microorganisms that break down chemical residues, thereby reducing the impact of pesticides and improving the environmental sustainability of agriculture. Technologies for utilising harmful emissions (chemicals, oil), polluting water and soil, and agricultural waste such as whey for producing food and feed protein products, particularly special enriched preparations, such as selenium yeast. For instance, the research by Bolatkhan et al. [6] demonstrates that cyanobacteria effectively reduce organic pollutant concentrations in wastewater, facilitating purification while also enabling the production of secondary raw materials, such as lipids for biodiesel or other bioproducts.

Rising prices for traditional energy sources (natural gas, oil, coal) and the threat of depletion have prompted scientists to turn to alternative energy sources. The role of biotechnology in creating economically renewable energy sources (alcohols, biogenic hydrocarbons, hydrogen) is extremely large [7]. These environmentally friendly fuels can be obtained by bioconversion of industrial and agricultural waste. Promising continuation of research to improve and implement the production of methane, ethanol, the creation of microorganisms (and enzymes) based on elements that efficiently produce electricity, as well as the organization of artificial photosynthesis, including biophotolysis of water from which energy-rich hydrogen and oxygen can be obtained.

The development of agricultural biotechnology at the present stage is aimed at solving such global problems as increasing soil fertility, yield and quality of agricultural products; reclamation of agricultural lands; improving the ecological situation, which contributes to the restoration of soil biocenosis; improving the quality of feed, etc. [8].

Environmental biotechnology also aims to create rational and harmless processes for humans and the environment by converting agricultural products into more valuable commodities [5]. To increase soil fertility, it is necessary to use organic fertilizers, composts, and liquid waste from livestock farms that are neutralized by methane fermentation.

Thus, **environmental biotechnology is divided into two areas** that meet the objectives of biotechnological processes:

1. Biotechnological methods used to protect the environment from industrial and domestic waste, degradation of various toxic compounds that have already entered the environment, etc.
2. Introduction of biotechnological methods during the implementation of low-waste (waste-free) industrial processes to fully utilise raw materials, including secondary raw materials (recycling), intensification of cultivation and use of renewable natural resources, etc. In addition to the production of new products, also products whose traditional methods of production are less environmentally friendly.

Biotechnological production's development on environmental safety principles is becoming increasingly important today.

The second direction of environmental biotechnology is based not only on indicators of the ecological efficiency of this technology but also on laying down effective economic levers of influence in the market of introducing "green technologies" that allow the expansion of their scope and make this process economically profitable. As noted by Matthews et al. [9], the sustainable development of emerging technologies, like synthetic biology, requires integrating environmental, economic, and social considerations through early-stage assessments and iterative learning to ensure their long-term viability and effectiveness.

An important area of environmental biotechnology is waste disposal from biotechnological industries. Advanced approaches, such as hybrid green environmental technologies that integrate biological systems and innovative treatments, are increasingly recognized as effective solutions for managing industrial wastewater and minimizing environmental impacts [10].

The problem of the relationship between biotechnology and ecology is the possibility of a negative impact on the biosphere of biotechnological processes and research in the field of biotechnology. It is necessary to consider the possibility of the adverse effects of bioconversion by-products, biosynthesis due to insufficient process control, out-of control biotechnological objects obtained by the latest biotechnology methods, biological contamination of adjacent media with mutagenic agents, etc.

The development and implementation of biotechnological production, as well as other types of technological processes, also leads to the emergence of industrial waste. Therefore, it is necessary not only to reduce the hazardousness of a pollutant and disinfect waste in the bioconversion process but also to look for ways to use by-products of processing (Fig. 1.3).

Thus, biotechnology and ecology can interact through both products and technology.

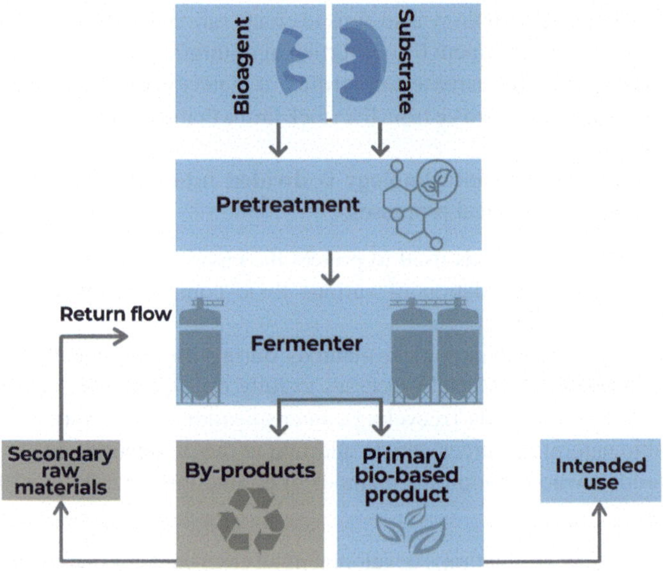

Fig. 1.3 Block diagram "Principles of low waste (no waste) in the development of biotechnological industries"

The primary purpose of introducing environmental biotechnology is to increase the level of greening in the national economy, which leads to a more harmonious relationship between society and nature.

Despite the challenges on the path to a bioeconomy, the constant demand for bioproducts and their timely delivery in an accelerated manner require a culture of knowledge exchange, digitalization, automation, and the development of flexible modular and capsule-based facilities to accelerate bioproduction [11].

1.2 Basic Elements of Biotechnological Processes

The main elements of biotechnological processes are the biological agent, substrate, equipment, process mode, and product. Biotechnological processes can employ various biological agents with different levels of organization, ranging from cellular to molecular.

A biological object is the basic structural unit of any biotechnological process.

Figure 1.4a and b shows diagrams of the interaction of the main components of the biotechnological process. Figure 1.4a—reflects the classic version of the structure, and Fig. 1.4b reflects the system "Biological object—substrate", which meets the principles of low waste and efficient resource use. For example, this scheme may describe a biotechnological process in which organic waste is both a substrate and a source of primary natural microbial grouping without introducing special groups of

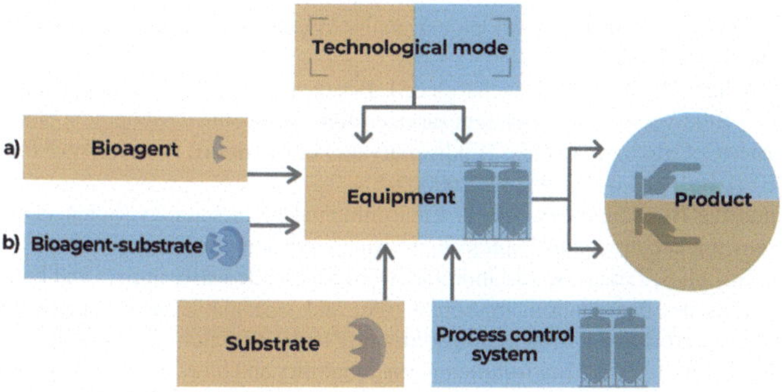

Fig. 1.4 Components of the biotechnological process: **a** classic structure (orange process flow). **b** Structure for low waste and efficient resource use (blue process flow)

microorganisms. This option also introduces a new element—the control system of the bioconversion process, which is now fundamentally important in the design and organization of biotechnological systems with self-regulation and management.

The main task in creating any biotechnological process is developing and optimising scientifically sound technology and equipment for it. In addition, many man-made low molecular weight compounds (pesticides, detergents, etc.) and high molecular weight polymers have proved to be stable and not entirely decomposed by natural microbial groups, i.e., the development of advanced technologies is required.

The emergence of new technologies aimed at extracting bioproducts using green technologies, food ingredients, or technological auxiliaries from marine sources has significant implications for the progress of food technology. One of the key challenges facing the marine food industry is the need to develop environmentally friendly production practices that minimize environmental impact and maximize resource utilization efficiency [12]. This involves optimizing extraction and processing technologies and adopting a holistic approach that considers the entire value chain from raw material sourcing to end-of-life disposal. Addressing social and economic aspects such as fair access to marine resources and fair labour practices is crucial for building a sustainable and inclusive food system. Furthermore, technological progress and innovation are key drivers of new opportunities and overcoming existing barriers. From bioprocessing technologies that extract valuable compounds from marine algae biomass to novel packaging materials made from polymers of marine origin, there exists a wide range of innovative solutions that require further exploration.

References

1. Timmis K, de Vos WM, Ramos JL, Vlaeminck SE, Prieto A, Danchin A, Verstraete W, de Lorenzo V, Lee SY, Brüssow H, Timmis JK, Singh BK (2017) The contribution of microbial

biotechnology to sustainable development goals. Microb Biotechnol 10(5):984–987. https://doi.org/10.1111/1751-7915.12818

2. Saleh M, Ashqar HI, Alary R, Bouchareb EM, Bouchareb R, Dizge N, Balakrishnan D (2024) Chapter 5—Biodiversity for ecosystem services and sustainable development goals. In: Singh K, Ribeiro MC, Calicioglu Ö (eds) Biodiversity and bioeconomy. Elsevier, pp 81–110. https://doi.org/10.1016/B978-0-323-95482-2.00005-5
3. Comprehensive Biotechnology (2011) Environmental biotechnology. ScienceDirect Topics. https://www.sciencedirect.com/topics/engineering/environmental-biotechnology
4. Gilani IE, Sayadi S, Zouari N, Al-Ghouti MA (2023) Plastic waste impact and biotechnology: exploring polymer degradation, microbial role, and sustainable development implications. Bioresour Technol Rep 24:101606. https://doi.org/10.1016/j.biteb.2023.101606
5. Arora NK, Fatima T (2024) Environmental sustainability and biotechnology: opportunities and challenges. Environ Sustain 7:115–119. https://doi.org/10.1007/s42398-024-00317-9
6. Bolatkhan K, Sadvakasova AK, Zayadan BK, Kakimova AB, Sarsekeyeva FK, Kossalbayev BD, Bozieva AM, Alwasel S, Allakhverdiev SI (2020) Prospects for the creation of a waste-free technology for wastewater treatment and utilization of carbon dioxide based on cyanobacteria for biodiesel production. J Biotechnol 324:162–170. https://doi.org/10.1016/j.jbiotec.2020.10.010
7. Auta HS, Aboyeji DO, Adelere IA, Babayi H, Bala JD, Fauziah SH, Aransiola SA (2024) Chapter 14—Biotechnology for bioenergy production: current status, challenges, and prospects. In: Maddela NR, Aransiola SA, Ezugwu CI, Eller LKW, Scalvenzi L, Meng F (eds) Microbial biotechnology for bioenergy. Elsevier, pp 277–296. ISBN: 9780443141126. https://doi.org/10.1016/B978-0-443-14112-6.00014-6
8. Reference Module in Food Science (2016) Agricultural biotechnology. https://www.sciencedirect.com/topics/agricultural-and-biological-sciences/agricultural-biotechnology
9. Matthews NE, Cizauskas CA, Layton DS (2019) Collaborating constructively for sustainable biotechnology. Sci Rep 9:19033. https://doi.org/10.1038/s41598-019-54331-7
10. Wang LK, Wang MHS, Shammas NK, Wang P (2022) Waste treatment in the pharmaceutical biotechnology industry using green environmental technologies. In: Wang LK, Wang MHS, Hung YT (eds) Waste treatment in the biotechnology, agricultural and food industries. Handbook of environmental engineering, vol. 26. Springer, Cham. https://doi.org/10.1007/978-3-031-03591-3_2
11. Pandey K, Pandey M, Kumar V et al (2024) Bioprocessing 4.0 in biomanufacturing: paving the way for sustainable bioeconomy. Syst Microbiol Biomanufact 4:407–424. https://doi.org/10.1007/s43393-023-00206-y
12. Augusto A, Lemos MFL, Silva SFJ (2024) Exploring marine-based food production: the challenges for a sustainable and fast biotechnology-based development. Appl Sci 14:8255. https://doi.org/10.3390/app14188255

Chapter 2
Biological Agents of Environmental Biotechnology

Abstract The biological agent is an active beginning in biotechnological processes and one of its most important elements. The range of biological agents is constantly expanding. The main ones include microorganisms, microbial associations, enzymes, fungi, and plants. Until now, the most important place is occupied by a traditional object—a microbial cell. It should be noted that environmental biotechnology is a low-waste technology with environmentally friendly technical implementation. They aim to clean the environment from various pollutants and produce environmentally friendly products with the possibility of recycling secondary resource flows. Therefore, the choice of biological agent here has a conceptual basis associated with the need for in-depth knowledge of physiology and biochemical processes of growth and metabolism of biological objects and the need for their sanitary and hygienic assessment.

2.1 Microorganisms

2.1.1 The Use of Different Physiological Groups of Microorganisms

Understanding microbial ecology is arguably one of modern ecology's most complex intellectual challenges [1]. ***Microbial cells*** with different chemical-technological properties can be isolated from natural sources and used further using traditional (selection and screening) and the latest methods (cellular and genetic engineering), which are significantly modified and improved. When choosing a biological agent and putting it into production, first of all, it is necessary to adhere to the principle of manufacturability of strains.

This means that the microbial cell, population or group of individuals must retain their basic physiological and biochemical properties during long-term fermentation. Industrial biological agents must also resist mutations, bacteriophages, and contamination by non-native microflora; be characterized by harmlessness to humans and the

Y. Chernysh et al., *Sustainable Biotechnology and Environmental Protection*, Green Energy and Technology, https://doi.org/10.1007/978-3-031-89436-7_2

environment; have no toxic by-products of metabolism and waste during cultivation; have high product yields and acceptable technical and economic indicators.

Today, many industrial microbial technologies are based on the use of heterotrophic organisms, and in the future, the decisive place among biological agents will be occupied by autotrophic microorganisms that do not require growth in deficient organic media, as well as **extremophiles**—organisms that develop in extreme conditions (thermophilic, alkalophilic, etc.).

The use of mixed microbial cultures and their natural associations has expanded in recent years.

In this kind of mixed culture, certain relationships are established between microorganisms based on the ecological principles of interaction in mixed populations. There are different ***types of such interaction:***

- neutrality—the practical lack of interaction between species.
- mutualism—both strains grow faster in a mixed culture than in the corresponding pure cultures.

Mutualism is a mutual exchange of growth factors. Profusely expressed mutualism is categorised as **symbiosis** (one microorganism cannot exist without another).

If one species produces substances, accelerating the growth of another species, it indicates that there is commensalism in the relationship between these species. The opposite of commensalism is **amensalism**, where one species produces a substance that inhibits the growth of another species.

Applying such ecological principles to microbial interactions is a foundation for developing effective biotechnological processes in industry [1].

Regarding microbes, four approaches have been used in their cultivation, in particular in selection work.

1. The selection of natural strains with valuable traits are found in the specific conditions of their habitat. This approach is based on microorganisms' natural course of events under changed conditions (natural selection).
2. Artificial selection of clones of microorganisms with useful features for humans arose based on the natural variability of parental forms.
3. Step selection is an effective method of artificial selection of forms using mutagens to increase the variability of test cultures and, among the latter, select the most promising.
4. Hybridization is a method of derivation of beneficial forms of microorganisms. There is an analogy with the method of crossing in higher eukaryotes.

2.1.2 Biophysical Bases of Microbial Culture Growth

The growth of a microbial culture is a complex biophysical process that includes the absorption of nutrients, metabolism, and cell division with biomass accumulation. The substrate for microbial growth can be organic or inorganic compounds that serve as sources of carbon, nitrogen, phosphorus, and other essential elements. The substrate concentration affects the growth rate of the culture, which is determined by kinetic parameters. Therefore, this subsection focuses on reviewing the fundamentals of microbial growth kinetics modelling.

Under the yield of cells (X), understand the difference between the maximum and initial biomass.

$$X = X_{max} - X_0. \tag{2.1}$$

This value is expressed in units of weight, more often—in grams. The ratio of cell yield to the amount of substrate consumed (X/S) is especially important. If both of these values are expressed in units of weight, the ratio X/S, called the economic coefficient, is denoted by Y:

$$Y = dX/dS, \tag{2.2}$$

where dX is the increase in biomass, corresponding to the substrate consumption by the amount of dS. The economic coefficient is more strictly determined by the limit to which this ratio approaches dS, which goes to zero. The economic factor is important because it expresses the quantitative needs of the body for food. For the first time, the economic coefficient was used by M. Rolen to express the nutritional needs of microscopic mycelial fungi ($Y = -\ dX/dS$). The "−" sign was introduced here because the values change in different directions. It is now accepted to use the value of Y, the inverse of that proposed by M. Rolen. By understanding the physiological characteristics of microorganisms and optimizing cultivation systems, researchers can unlock new opportunities for the sustainable production of biofuels, chemicals, and other valuable compounds. A thorough understanding of the physiological characteristics of microorganisms, combined with the optimization of cultivation systems, enables the development of sustainable and efficient processes for producing biofuels, chemicals, and other high-value compounds [2].

If the yield (in grams) is attributed to moles of the consumed substrate, the economic coefficient, which is called, in this case, the molar economic coefficient, is denoted by Y_m. The molar economic coefficient Y_m allows us to link the yield of cells with the amount of ATP (macroergic equivalents) obtained from any energy source (any substrate). Y_{ATP}—energy factor expressed in grams of cell mass per 1 mol of ATP. This factor can be calculated if the path of cleavage of this substrate and the yield of ATP as a result of this cleavage are known. The rate of consumption of the substrate by the culture at this time is expressed by the ratio

$$dS/dT = qX \tag{2.3}$$

where X is biomass, and the coefficient q is known as the metabolic coefficient or the specific rate of metabolism.

The metabolic coefficient is similar to the enzymatic activity [3].

The metabolic coefficient can also be expressed in terms of economic coefficient (Y) and specific growth rate (μ) and presented as follows:

$$q = \mu / Y \quad (2.4)$$

The parameter μ, which means the growth rate of a unit of biomass (*1/X*) (*dX/dt*), is called the specific growth rate and is measured in units inverted in time (*1/t*). This parameter is similar to compound interest. For example, a specific growth rate of 0.1 h is equivalent to a rate of 10% for 1 h.

This parameter can be calculated considering that

$$dX = \mu \cdot Xdt \quad (2.5)$$

where *dX/dt* is the growth rate; *X*—biomass; μ is the coefficient of proportionality ("specific growth rate"),

From here

$$dX/dt = \mu X \text{ or } \mu = dX/dt \cdot 1/X \quad (2.6)$$

If μ is constant, then the integration of the equation will look like this:

$$LnX = LnX_0 + \mu t \quad (2.7)$$

where X_0—biomass at the initial time $t = 0$.

The growth of the cell population subject to this law is called exponential or logarithmic growth. The graph of *lnX versus* time will look like a straight line with a slope of μ. The specific growth rate is one of the main parameters that characterize the physiological state of the biological agent; a number of other parameters can be expressed through this indicator.

To calculate the time of cell generation, you can use the equation, taking into account the geometric progression of growth:

$$N = N_0 \cdot 2^n, \text{ hence } lgN = lgN_0 + nlg2 \quad (2.8)$$

where N is the number of cells.

Hence, the number of cell divisions (n) will be

$$n = lgN - lgN_0/lg2 \quad (2.9)$$

If we plot this process on a graph, we obtain an exponential growth curve in which lgN/t is transformed into a straight line in semi-logarithmic coordinates. In practice, exponential growth can be approximated by two straight lines at three points (Fig. 2.1) and perceived as a threshold event with a sudden transition to saturation after reaching the maximum density.

The initial phase of the exponent is usually perceived as a lag phase, partly due to the threshold of recognition of the beginning of growth by turbidity by $lgN_t = 5$ or 6. It is necessary to distinguish the true lag phase from this imaginary lag phase if there is no reproduction and there is a cell preparation before the beginning of reproduction

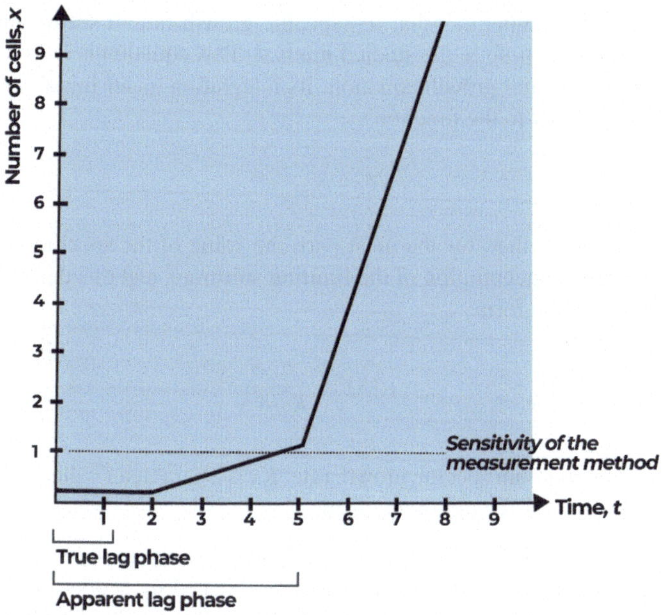

Fig. 2.1 Approximation of exponential growth by straight lines

by processes of initiations in the middle of a cell. The lag phase is determined by a time frame cut off on a semi-logarithmic graph of the direct number of living cells. As we approach the limit, growth slows. Usually, the maximum amount is determined using the nutrient substrate, but there may be other causes, such as poisoning by metabolic products [4].

There are also more subtle mechanisms of population density regulation with the use of signalling molecules, such as the increase in the concentration in the environment perceived by the genetic apparatus of bacteria. After reaching the maximum possible density under these conditions, the bacteria enter a stationary phase, characterized by a state where cell division almost completely stops. The products of metabolism and the substrate are limited, while the number of cells remains high, and accordingly, cell division nearly ceases.

There are two features of the empirical dependence of the growth of cultures of microorganisms [5].

1. The rate of change in the number of microorganisms in the mode of its growth (in the exponential phase) is linearly related to the concentration of cells in the system:

$$\frac{dN}{dt} = \mu N \tag{2.10}$$

where N is the number of cells; μ—specific growth rate. It seems that μ does not depend on the time in the studied interval. This equation is in integral form and is an exponential growth equation. Its integration under initial conditions $t = 0$, $N = N_0$ leads to the function

$$N = N_0 \cdot e^{\mu t} \tag{2.11}$$

2. It was determined that, for the most part, the value of the specific growth rate depends on the concentration of the limiting substrate, and this dependence can be presented in the form

$$\mu(S) = \frac{\mu_m S}{K_s + S} \tag{2.12}$$

where μ_m—maximum specific growth rate; K_S is a parameter called the substrate affinity constant for the microorganism.

There is a hyperbolic relationship (Fig. 2.2) between the specific growth rate μ and the substrate concentration *S*, which was established by Monod, and therefore Eq. (2.12) is called the Monod equation. This dependence is clearly manifested during the periodic cultivation process [5].

A fundamental feature of the kinetics of microbial populations is the dependence of the culture's growth rate on the concentration of one or more of the most important components of the environment that provide the biosynthetic basis of metabolism. These components are called ***limiting substrates*** and, to some extent, regulate the growth rate of the population.

There are *two approaches to describing the dependence of the growth rate of* cultures of microorganisms on the concentration of the limiting substrate—macroscopic and microscopic [6, 7].

In the macroscopic approach, the growth of the culture of microorganisms is presented as an autocatalytic process. It quite accurately describes the fundamental

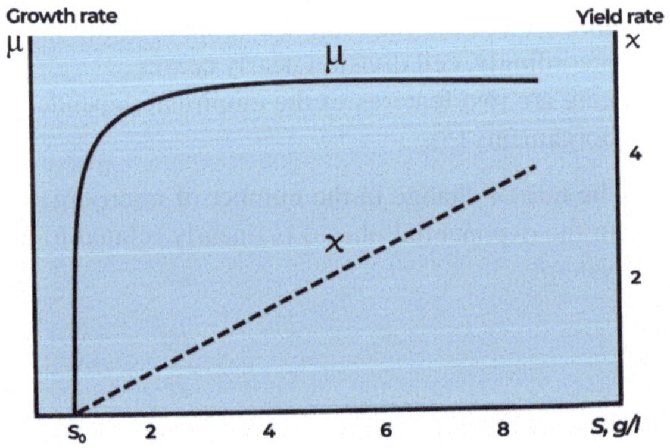

Fig. 2.2 Dependence of growth rate and yield from the initial concentration of the substrate

kinetic features of microbial growth—the exponential nature of the process and the dependence of the specific growth rate on the substrate concentration. The equation linking the growth rate of cultures of microorganisms with the substrate concentration (Monod equation) is empirical [8]. Let us imagine a simple kinetic scheme of the autocatalytic reaction with the decomposition of catalytic centers:

$$S + N \rightarrow^{\kappa} 2N + P \tag{2.13}$$

where N is the number of centers that have doubled when interacting with the substrate; P—product.

The number of centers with which a directly proportional relationship relates the substrate interacts and the number of cells of microorganisms because it is assumed that the cells, on average, contain the same number of centres at which the transformation of the substrate. In the framework of kinetic approximation, the frequency of cell division acts can be denoted by the constant k. Due to interaction with cell components, Substrate S is transformed to form product P.

The equations will describe the dynamics of the change in the system of concentration of components

$$\frac{dN}{dt} = kSN \tag{2.14}$$

$$\frac{dP}{dt} = \frac{kSN}{Y_p} \tag{2.15}$$

where Y_p is the coefficient of proportionality that connects the amount of biomass formed and the product of enzymatic transformation:

$$N - N_0 = Y_p(P - P_0) \tag{2.16}$$

If the concentration of substance S in the system is large enough and changes insignificantly during the process, it can be considered a constant value ($S = S_0$). In this case, Eq. (2.14) can be integrated with respect to the number of cells N. If we use the initial condition ($t = 0, N = N_0$), the integration of Eq. (2.14) is reduced to an exponential function

$$N(t) = N_0 e^{kS_0 t} \tag{2.17}$$

Substituting Eq. (2.17) into differential Eq. (2.14) and its subsequent integration (assuming that $P_0 = 0$) lead to the following dependence on the time of concentration of the product:

$$P(t) = \int_0^t \frac{kS_0 N_0}{Y_0} e^{kSt} dt = \frac{N_0}{Y_p}\left(e^{kS_0 t} - 1\right) \tag{2.18}$$

At a reaction time close to zero $\left(t\langle\langle\frac{1}{kS_0}\right)$, the concentration of the product is close to zero, at a large reaction time, there should be an exponential increase in the concentration of the product:

$$P(t) = \frac{N_0}{Y_p}e^{kS_0 t} \tag{2.19}$$

Equations (2.17), (2.19) and, accordingly, the kinetic scheme (2.13) describe an important feature of the growth kinetics of the microbial population—the number of cells in the system increases exponentially over time. The specific growth rate μ (S), which is an exponent in Eqs. (2.17) and (2.19), is a linear function of the substrate concentration:

$$\mu(S) = kS_0 \tag{2.20}$$

Equation (2.20), in contrast to the Monod equation, leads to a linear unlimited increase in the specific growth rate of microorganisms with increasing substrate concentration. This contradicts the experimental data and requires additional complication of a simple kinetic scheme of population growth when detailing it, taking into account two irreversible stages of the process:

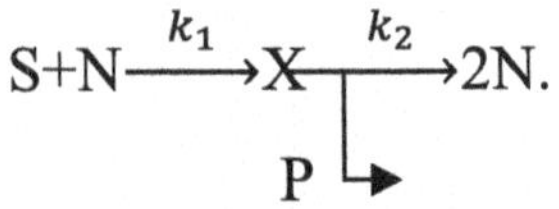

(2.21)

The substrate, interacting with the catalytic centers inside the cell (the number of centers N is proportional to the number of cells in the population), leads to the formation of metabolites that allow cell division and doubling of the centers with which the substrate interacts. Then we obtain the hyperbolic function of the substrate concentration corresponding to the Monod equation:

$$\mu(S) = \frac{k_2 \cdot S_0}{\frac{k_2}{k_1 + S_0}} \tag{2.22}$$

The value of μm has the meaning of the rate constant of the doubling process of the centers k_2 with which the substrate interacts. K_s in the Monod equation is the ratio of the constants of the two stages of the process k_2/k_1 [9].

The methodology for constructing kinetic equations of microbial growth can be constructed by another approach—***microscopic***, i.e. based on chemical-kinetic representations of the doubling time of cellular material, based on the assumption that some limiting biochemical process determines time. Provided that this time can be calculated, population growth in the exponential phase will be described by the equation:

$$M = M_0 \cdot 2^{t/\tau} \tag{2.23}$$

where τ is the characteristic time of the biochemical cycle, the doubling time, M_0—initial concentration of cells.

This fairly general approach allows us to analyze the growth kinetics of the cell population if it is possible to quantify τ. It is assumed that the action of the enzyme (or enzyme system) that transforms the limiting substrate and the process of DNA replication are the slowest stages. This is characteristic of actively growing microorganisms [10].

2.1.3 The Use of Microorganisms in the Processes of Environmental Biotechnology

Biopesticides

Bacteria, fungi, and viruses are increasingly used as industrial biopesticides. The production technology of these agents is very different, and so is the diverse nature and physiological characteristics of microorganisms-producers. However, there are a number of universal requirements for biopesticides, the main ones of which are selectivity and high efficiency, safety for humans and beneficial representatives of flora and fauna, long-term preservation and ease of use, good wettability and tack [11]. Unicellular fungi are well preserved in the form of spores and produce a variety of biologically active substances that increase their pathogenicity [12]. However, fungal drugs are not used widely enough. This is due, firstly, to certain technological difficulties that arise during their cultivation, and secondly, due to strict requirements for environmental factors (high activity of fungal drugs is detected only in conditions of high and stable humidity).

Production of Azotobacterin

Azotobacterin is a bacterial fertilizer containing the freely existing soil microorganism *Azotobacter chroococcum*, capable of fixing up to 20 mg of atmospheric nitrogen per 1 g of sugar used. Bacteria introduced into the soil release biologically active substances (nicotinic and pantothenic acids, pyridoxine, biotin, heteroauxin, gibberellin, etc.). These substances stimulate plant growth. In addition, the fungicidal substances produced by Azotobacter from the anisomycin group inhibit the development of some undesirable microscopic fungi in the plant's rhizosphere.

It is established that during nitrogen fixation, its reduction occurs on the same enzyme complex synthesized by azotobacter, and only the final product (ammonia) is separated from the enzyme. Nitrogenase nitrogen fixation system is a multi-enzyme complex containing non-gene-related iron, molybdenum and SH-groups.

It is recommended to use azotobacterin only on soils containing phosphorus and microelements. Azotobacterin is used to bacterize seeds, seedlings, and composts [13].

Providing Plants with Phosphates

Phosphate ions in the soil are not known to be very mobile, so phosphorus deficiency often occurs around the root zone of plants. Vesicular–arbuscular mycorrhiza (VA) plays a significant role in soil fertility, as it promotes the uptake of phosphates by plants from the soil. Endo- and ectomycorrhizae are special structures that form inside or around the small roots of plants due to infection by soil non-pathogenic fungi.

Symbiotic relationships that occur between fungi and plants are beneficial to the host plant. VA mycorrhiza, formed by the fungus phycomycete of the family Endogonaceae, occurs quite often in most soils of almost all climatic zones. This mycorrhiza is characteristic of most angiosperms, many gymnosperms, some ferns and liverworts. Mycorrhiza VA is found in most important species of cultivated plants. Mycorrhizal hyphae, which grow from the mycelium and spread far beyond the root system, transfer phosphate ions from the zones of their presence to the host cells. VA has the greatest effect on plants with a weak root system. Endophytes in the soil require their inoculation to reproduce. However, the reproduction of fungi occurs only in the presence of the host plant. Growing plants on the appropriate line is the only effective way to obtain large amounts of endophytes. The inoculum is a mixture of mycelial roots and spores. A few grams of crude inoculum obtained from the pot culture of the host plant is added to the environment or placed near the young roots so that before transplanting, the plant in the soil has time to form a fairly strong mycorrhiza. This method is effective when planting forests and citrus, but it is not used for inoculation in the field because the drug requires a lot (2–3 tons of crude inoculum per 1 ha). It is not possible to obtain such amounts of VA inoculum [14].

The technology of obtaining the bio-preparation phosphorobacterin is in many respects similar to the technology of obtaining dry nitragin and azotobacterin. Growing *Bacillus megaterium* is carried out in a controlled deep culture to the spore formation stage. The process is carried out under strictly sterile conditions, as many production strains are sensitive to bacteriophages [15].

2.2 Microbial Associations and Consortia

The association of microorganisms, a grouping of microorganisms, is constantly observed and (or) develops interdependently. ***The basis for the formation of the association is***: the sequence of decomposition of the substrate by different microorganisms (metabolic association); exchange between them of growth factors, for example, vitamins (protocooperation); removal of toxic metabolic products (amensalism); energy substrate exchange (syntrophic associations). Associations of microorganisms can turn into morphologically formed groups of co-developing bacteria, such as sulfate-reducing and green phototrophic bacteria. In a broad sense, the association of microorganisms—a microbial population that is constantly observed together in a particular substrate, and the physiological relationship of its components is not always clear; for example, the association of microorganisms in meadows, podzolic and other soils [16].

Syntrophism is a special case of symbiotic cooperation between metabolically different types of bacteria, which depend on each other during the destruction of substrates.

The term "consortium" is used to describe various communities of microorganisms. A footnote to Rule 31 of the International Bacterial Nomenclature Code (1978) states: "**A consortium** is an aggregate or association of two or more organisms". Thus, this concept includes such groups of microorganisms as an association and mixed culture [17].

Compared to monocultures, microbial associations can consume complex, heterogeneous substrates, mineralize complex organic compounds, have an increased ability to biotransform, increased resistance to adverse environmental factors and toxic substances, increased productivity and the ability to exchange genetic information between species [18].

The main areas of application of mixed cultures are food production, environmental protection, biodegradation and assimilation of complex substrates.

It should be noted that today, in the field of biotechnology and environmental protection development, an important place is occupied by associations of microorganisms, which are distributors of organic pollutants. They are used in aerobic and anaerobic wastewater treatment, in the processing of household waste and organic wastewater, in the bioremediation of contaminated soils, etc. Therefore, it is very important to understand the biochemical mechanisms of microbial groups and analyse their use in environmental biotechnology.

Competitive relations arise between species capable of using a single substrate. There is no such relationship between members of the group using different substrates. Competition between them may be a secondary factor, but the main type of interaction in the microbial group—trophic—does not give rise to competitive relations. Thus, competition occurs within one physiological group. Success in the competition implies, by default, the most achievable biomass of this species, usually expressed in terms of the number of individuals N. This criterion is sufficient within certain limits, but there may be three other less obvious criteria, such as the survival of individuals, species longevity, or persistence. Obviously, the number of individuals cannot increase indefinitely but reaches a certain limit due to the capacity of the environment K. If the growth rate depends on the already accumulated number of individuals and is described by the logistic equation of Verhulst [19]:

$$\frac{dN}{dt} = r \cdot N \cdot \left(1 - \frac{N}{K}\right) \tag{2.24}$$

where r is the coefficient of exponential growth; dt—is the time of accumulation of microorganisms in the soil.

Imagine two organisms, A and B, using the same substrate. For them, we can represent two curves of dependence μ on s. Let A μ_m be smaller than B, but let K_s be smaller than B. Then, there is a range of small concentrations s at which A, which grows slowly, benefits from B, which grows rapidly. This is the ratio of the characteristics of two competing organisms, called "*rules of intersecting curves*": *at low substrate concentration wins an organism with high substrate affinity and low* K_s, *and at high concentration—an organism with high* μ_{max}. Accordingly, they are referred to as **oligotrophs and copyotrophs**. In fact, both organisms grow in the same vessel or habitat, but one grows larger. The minimum concentration of a substrate at which transport systems of an organism are already inefficient is important. This value often determines the residual concentration of the substrate in nature after the microorganisms reach the maximum density. For example, the threshold concentrations of H_2 in the environment differ significantly for different groups of anaerobic hydrogenotrophic organisms [20].

The lowest affinity for hydrogen in homoacetogenic organisms is approximately 1000 ppm (parts per million, 10^{-6} by volume) in the gas phase, followed by methanogens with a threshold of 5–100 ppm; in sulfate reducers, this value is less than 1 ppm, and in bacteria that reduce iron, even lower. Since the concentration of the reaction product affects its thermodynamic yield, the growth of anaerobes that form hydrogen depends on the final level of H_2 set in the environment. In competition, survival is no less important than the rate of growth, especially at the minimum concentration of the substrate: not only the organism whose number is greater at the moment wins, but also the one whose extinction rate is lower.

However, moving to natural conditions, we can come across a number of additional considerations that limit the absolutization of the rule of crossed curves. First, despite the slower growth rate in binary culture, *A* still grows and will be a certain part of the population. Removal of *A* from the culture can only occur with the same probability of death for *A* and *B*, for example, due to leaching from the flow culture. Still, the probability of death is determined by factors other than the growth rate. Second, the *successful colonization of the substrate depends on the initial number of cells* N_0 and with a large sowing that grows relatively slowly, *A* may dominate. This possibility arises due to a large number of stable cells, such as spores, in the environment and depends on the speed of their lag phase, i.e. awakening to active growth. Finally, thirdly, μ_m characterizes certain environmental conditions and varies widely depending on temperature, pH, and salinity. Due to the use of the same substrate in different conditions, different organisms are preferred, and the number of potential opportunities for species diversity in performing the same function in the group increases many times over [21].

When considering competition based on growth kinetics, it is implied that we are talking about a substrate as an electron donor. However, in addition to the donor, the body needs an acceptor. The solubility of oxygen in water is very low and is less than 10 mg/l. With oxygen restriction, there is competition between aerobic organisms and microaerobic (microaerophiles), which develop at oxygen concentrations of approximately 1 mg/l. Hence, competition between aerobic organisms, which use different substrates, can be fierce.

Restriction of growth by "limiting factors" may belong not only to the energy substrate of catabolism but also to the substrates of anabolism, which are united by the concept of necessary *growth factors*. These may include amino acids, vitamins, and minerals. For all these limiting factors in chemostatic culture (with the same probability of death), the rule was derived that the number of species should equal the number of limiting factors [22].

Thus, based on the high rate of reproduction of microorganisms as organisms with a short life cycle, we can expect that in nature, most organisms are in a state of starvation or limitation of the substrate and, accordingly, the rate of their reproduction is minimal. The high rate of bacterial growth observed in laboratory cultures is a myth, and it is dangerous due to its resistance; it can only be a potentially high reproduction rate. The physiological state of organisms in nature determines their ability to reproduce. Based on molecular methods for identifying the species of microorganisms and their number in nature, it was found that a significant part of the population belongs to the so-called uncultivated forms, i.e. those that do not germinate when sown in laboratory media.

Thus, for bacteria, the main expression of competition is the kinetics of growth and death. The growth rate depends on the concentration of the substrate, and as it depletes, there should be a change in the dominant species.

In the trophic chain, where the metabolic product of one organism A is a substrate for organism B with other possibilities of exchange, there is a trophic chain

$$s - A - a - B - b - C - p \tag{2.25}$$

where s is the substrate at the entrance to the food chain; p is the output product.

The number of organisms tends to the maximum, and the concentration of intermediate products (a, b)—to a minimum. As a result, the organism first adapts to the decreasing substrate concentration by incorporating increasingly efficient substrate capture systems and increasing affinity for it by changing transport systems. If the substrate concentration reaches the minimum concentration threshold for this organism, it can be changed to another species. great kinship. A classic example is the change in the above-described methanogenic grouping of acetoclastic organisms: at the beginning, at a high concentration of acetate, it is dominated by large methanosarcin with $K_s = 5$ μm and a relatively high growth rate, and when the concentration decreases it is replaced by *filamentous methanotrix* with $K_s = 0.5$–1 μM and five times lower growth rate. As a result, for a long time, the group's existence should be dominated by species of organisms with a high affinity for the substrate, capable of developing at a very low concentration, which is referred to as the threshold. At the same time, the largest share of available energy is spent on support energy at low substrate concentrations, which is not used to increase the number of organisms [21].

2.2.1 Redox Processes in the Microbiological System

Microorganisms receive energy for their vital activity due to redox reactions that transfer electrons from the donor to the acceptor. Each individual organism is an electrochemical element, carrying out two half-reactions separated by a membrane.

For inorganic compounds, the thermodynamic sequence of redox potentials is clearly adhered to as the preferred acceptors; the least advantageous thermodynamic acceptors are used last. Therefore, *developing secondary anaerobes is possible only in the thermodynamic stability of the reduced reaction product in the Eh-pU coordinates* [23].

The main energy sources for living beings are solar radiation carbohydrates; only for prokaryotes—H_2, CO_2, NH_3, N_2 and its inorganic derivatives (H_2S, $Na_2S_2O_3$, SO_4^{2-}), and Fe^{2+}. Bacteriorhodopsin, the redox chain of phototrophs, and the respiratory chain are among the primary energy generators in which light and chemical energy are converted into a form accessible to cells. In chemotrophic or chemosynthetic species, the set of redox systems is broader. Thus, secondary anaerobes' most important electron donors are hydrogen and HCOOH, acetate and ethanol, and volatile fatty acids (VFA). These compounds are contained in major trophic routes. *The sequence of using electron acceptors involves:* (a) reduction of Fe (III); (b) reduction of nitrates, denitrification, anaerobic formation of ammonium; (c) reduction of sulfur compounds in hydrogen sulfide, or sulfidogenesis; (d) reduction of carbon dioxide to methane, or methanogenesis. The latter process does not require external oxidant sources, as carbon dioxide is formed during fermentation. Each of these processes is observed in nature under certain conditions depending on the arrival of the acceptor. Their sequence is clearly visible in time, for example, when the soil is flooded, and in space, for example, in the profile of ocean silt precipitation [24].

Consider the example of half-reactions to construct the balance equation of the general reaction of microbial synthesis [25, 26].

General reaction

$$R = R_d + f_e R_a + f_s R_c \tag{2.26}$$

where f_e *is* the proportion of electron donors used to generate energy; f_s—is the proportion of electron donors used for cell synthesis; at $f_e + f_s = 1$.

Electron donor half-reactions (R_d)

1. *Electron donors*

   (a) organic liquid

$$\frac{1}{e}C_aH_bO_cN_d + \frac{2a-c}{b}H_2O = \frac{a}{c}CO_2 + \frac{d}{e}NH_3 + H^+ + e^- \tag{2.27}$$

   where e = 4a + b − 2c − 3d;

   (b) inorganic liquid

$$\frac{1}{8}NH_4^+ + \frac{3}{8}H_2O = \frac{1}{8}NO_3^- + \frac{5}{4}H^+ + e^- \quad (2.28)$$

$$\frac{1}{16}H_2S + \frac{1}{16}HS^- + \frac{1}{2}H_2O = \frac{1}{8}SO_4^{2-} + \frac{19}{16}H^+ + e^- \quad (2.29)$$

$$Fe^{2+} = Fe^{3+} + e^- \quad (2.30)$$

Half-Reactions of the electron acceptor (R_a)

2. *Electron acceptors*

Aerobic process

$$\frac{1}{4}O_2 + H^+ + e^- = \frac{1}{2}H_2O \quad (2.31)$$

Anaerobic process

- denitrification

$$\frac{1}{6}NO_3^- + H^+ + \frac{5}{6}e^- = \frac{1}{12}N_2 + \frac{1}{2}H_2O \quad (2.32)$$

- sulfate respiration

$$\frac{1}{8}SO_4^{2-} + \frac{19}{16}H^+ + e^- = \frac{1}{16}H_2S + \frac{1}{16}HS^- + \frac{1}{2}H_2O \quad (2.33)$$

- carbonate respiration

$$\frac{1}{8}CO_2 + H^+ + e^- = \frac{1}{8}CH_4 + \frac{1}{4}H_2O \quad (2.34)$$

Consider **the half-reactions of microbial cell synthesis (*R_c*).**

Source of ammonium N:

$$\frac{1}{4}CO_2 + \frac{1}{20}NH_3 + H^+ + e^- = \frac{1}{20}C_5H_7O_2N + \frac{2}{5}H_2O. \quad (2.35)$$

Source of nitrate N:

$$\frac{5}{28}CO_2 + \frac{1}{28}NO_3 + \frac{29}{28}H^+ + e^- = \frac{1}{28}C_5H_7O_2N + \frac{11}{28}H_2O. \quad (2.36)$$

2.2.2 Energy of the Microbiological System

In fact, in the energy of microorganisms we are dealing with an electrochemical reaction. Microbiologists need thermodynamic calculations in EH–PH coordinates to determine the metabolism of lithotrophic organisms.

The ability to use the energy of a chemical reaction to grow organisms is determined by the change in free energy, ΔG (that is, the energy that can be converted to work at constant pressure and temperature).

The concept of physical chemistry

Change of free energy in the reaction

$$A + B = C + D \tag{2.37}$$

can be calculated through the equilibrium constant of the reaction as

$$\Delta G = \Delta G^o + 2.303R \cdot T \cdot \lg K = \Delta G^o + \log \frac{[C][D]}{[A][B]} \tag{2.38}$$

where R is the gas constant equal to 8314 J/(mol K); T is the absolute temperature in Kelvin; K is the equilibrium constant of the reaction; $[A]$, $[B]$, $[C]$, $[D]$—concentrations of substrates (A, B) and products (C, D) of the reaction; ΔG^o is standard change of free energy, calculated for standard reaction conditions [27]: temperature 25 °C, 298 K, pressure 1 atm and 1 M concentration of all reactants,

$$\Delta G^o = 2.303R \cdot T \cdot \lg K, \tag{2.39}$$

and this value is a constant characteristic of the reaction. To denote the standard free energy at pH = 7, which corresponds to physiological conditions, use the symbol ΔG^o.

The most important consequence of this equation is the dependence of the growth of microorganisms on the removal of reaction products, which is achieved by their interaction with microorganisms that consume these products. Another option is to remove the products in the form of gases or the formation of insoluble minerals. These patterns are applicable not only to one reaction, but also to the chain of reactions carried out by the group.

The change in standard free energy can be substances and the free energy of the reaction products under standard conditions.

Standard change of free energy ΔG^o can be expressed as:

$$\Delta G^o = \sum G^o_{product} - \sum G^o_{reagent}. \tag{2.40}$$

Microorganisms develop in the field of stability of the reaction product and instability (metastability) of the substrate [28].

Just as energy-giving reactions are combined within the cell of a microorganism in a chemical system of combined redox reactions carried out by a microbial group, it is possible to combine them in a sum that gives an energy gain sufficient for the group to exist. However, unlike a cell with an energy carrier, in a group, the conditions must provide a gain sufficient for the existence of each individual group or species of organisms.

A fundamental discovery in the energy of interaction between organisms was the discovery of catabolic syntrophy by M. Brian and E. Volyn in the 70 s of the twentieth century. Based on interspecific hydrogen transfer [29].

Catabolic syntrophy is the realization of a catabolic reaction only in the case of the interaction of two organisms. The phenomenon of catabolic syntrophy means that the grouping of microorganisms can and should be considered a biological unity due to the thermodynamic capabilities of its components [30].

There are thermodynamically conditioned feedback and regulatory connections between groups of microorganisms. For example, the decomposition reaction of VFA to acetate and hydrogen is not thermodynamically advantageous. It can ensure the growth of microorganisms only at very low concentrations of reaction products, i.e. rapid and complete removal of H_2. The function of removing H_2 is performed by hydrogenotrophs (primarily hydrogenotrophic methanogens and in the presence of electron acceptors SO_4^{2-} sulfate-reducing bacteria). Organisms capable of development provided that H_2 is released from the reaction sphere, are called syntrophs and constitute a special functional group within the community. From unfermented compounds, they form acetate and hydrogen. Syntrophs develop in conjunction with H_2-used and anaerobic organisms. Among the H_2-used bacteria, two groups are important: lithotrophic methanogens and lithotrophic sulfidogens. Several cultures have been described in which anaerobic oxidation of the substrate is due to the presence of an H_2-used component.

Thus, feedback is fundamentally important in the group and acts as a metabolic whole. Consider biochemical models of trophic relationships for some aerobic and anaerobic groups and bacterial oxidation filters (gasotrophs). These microbial groups are actively used in the processes of biological wastewater treatment. In Crognale et al. [31], the contribution of indigenous groundwater bacteria to As(III) oxidation in biofilters under conditions suitable for household wastewater treatment systems was thoroughly investigated. The study explored the influence of various experimental conditions, including different bed materials, flow rates, arsenite (As(III)) concentrations, As(III): As(V) ratios, and biofilter volumes, on the microbially-mediated arsenite (As(III))-oxidation process in terms of its efficiency and oxidation rate.

2.2.3 *Aerobic Grouping of Microorganisms*

The central place in the aerobic group is occupied by primary biological agents—oxygen photoautotrophs. During the day, they are a source of O_2 and excreted organic matter, which can be used by concomitant organotrophs and for vascular plants—and the inhabitants of rhizoplanes that consume root secretions. At night, photoautotrophs use spare substances for respiration. In those benthic groups of cyanobacteria or algae, which are at night in the anoxic zone, dark metabolism can be provided by the fermentation of spare carbohydrates in the form of polyglucose and cyanobacteria—and fermentation of arginine. The main product of primary producers is the mortar of photoautotrophs, which contains two categories of substances: easily degradable spare substances and components of the cytoplasm, mainly soluble and insoluble structural components, primarily cell wall, which comes in the form of precipitation.

Aerobic organotrophs are highly diverse and have attracted more attention at the level of pure cultures. The system of trophic connections is characterized by the structure of convergent food routes in the form of an inverted "fan", where each substance has its specific action. The "fan" coincides with CO_2 as the final product, and the total effect of aerobes is determined by respiration: formed CO_2 and/or consumed O_2. In addition to specialization by substrates, it is possible to distinguish groups of zymogens-copiotrophs, hydrolytics, dissipotrophs, and autochthons in the trophic system. Using substances excreted by oxygen phototrophs leads to temporary exposure of organotrophs under hyperoxia conditions. Therefore, aerobes in this group must effectively protect against the toxic effects of O_2, which light enhances [32].

Dissolved substances are the substrate of the zymogenic microflora represented by copyotrophs. Protein compounds are decomposed by proteolytic decay with the participation of ammonifying organisms; ammonia enters the nitrifiers. In nature, the stages of degradation can be observed in the self-purification of watercourses, in which there are zones of saprobity from heavily polluted to purified water. Aerated treatment plants are a model of intensive decomposition of achievable organic matter. The critical factor is oxygen availability: at a high organic matter content, its supply by diffusion becomes insufficient and anaerobic organisms predominate. A practical method of determining respiration due to solutes is determining the "biochemical oxygen demand" (BOD). Usually, the consumption of O2 for 5 days (BOD_5) in the studied natural water or after its dilution with aerated water is determined. BOD is a reliable measure of determining the available organic matter [33].

Dissolved carbon dioxide and oxygen concentrations can characterize the ratio of the rates of production and degradation processes. For reservoirs, this ratio can be expressed as a function of pH and concentration of dissolved O_2 [34].

At constant illumination and concentration of biomass of phototrophs and heterotrophs in the first approximation of the speed of the processes of photosynthesis and destruction can be described by the equations:

$$v_{phot} = k_1[CO_2] \tag{2.41}$$

$$v_{dest} = k_2[CO_2] \tag{2.42}$$

where k_1 and k_2 are the rate constants of photosynthesis and destruction.

Then

$$\frac{v_{phot}}{v_{dest}} = \frac{k_1[CO_2]}{k_2[CO_2]} \tag{2.43}$$

The release of CO_2 during respiration causes a decrease in pH in natural waters due to the formation of H_2CO_3. The absorption of CO_2 during photosynthesis leads to an increase in pH.

Thus, in reservoirs, the ratio of production rates and destruction can be expressed as a dependence

$$\frac{v_{phot}}{v_{dest}} = f(pH, [O_2]) \tag{2.44}$$

Primary production and destruction are determined by the amount of O_2 released by photosynthetics at the water temperature of the reservoir, and the amount of O_2 absorbed under the same temperature conditions.

Volatile organic compounds (VOCs), such as terpenes released by plants and toxic to bacteria, play a special role among substances excreted by plants.

Insoluble substances are subject to hydrolysis by specialized groups of organisms. In reservoirs, suspended organic matter significantly settles to the bottom and enters the anaerobic zone due to the slow diffusion of O_2 [35].

In the system of aerobic destruction, the characteristic product is residual substances, such as lignin and compounds that are difficult to decompose, which give rise to humic substances. Lignin decomposition is especially important, as it is carried out only by fungi using a peroxide mechanism, which depends on oxygen availability. The phenolic compounds formed are used by bacteria. The grouping of fungi as the most important destruction of woody plant mass acts on the main trophic routes in forest ecosystems. Fungi are not typical for aquatic environments and are considered secondary settlers.

The regional specificity of mVOCs, soil conditions, and specificities of volatilized microbial communities necessitate the efficiency of "Green Revolution 2.0" [36]. Thus, the usefulness of mVOCs and their numerous advantages for effective plant growth promotion and eco-friendly applications in sustainable agriculture are underscored. Environmental factors, such as microbial growth conditions, microbial community composition, nutrient and oxygen availability, temperature, and pH, influence volatile organic compound (VOC) production. Commercial applications of these volatiles are significantly limited compared to their economic implications. Therefore, limitations in applying volatiles can be overcome by employing economically efficient technological achievements and eco-friendly management strategies. Despite extensive research on the efficacy of various VOCs in plant–microbe interactions, the precise mechanisms remain unknown. This highlights the importance of further research into VOCs and plant–microbe interactions.

2.2.4 Anaerobic Grouping of Microorganisms

The process of anaerobic fermentation consists of several stages or phases. It should be noted that the first phase is usually hydrolytic/acidogenic, as complex polymeric organic compounds cannot be converted to methanogens, acetogens or sulfate-reducing bacteria. But the simple compounds formed by hydrolytics (H_2, CO_2, etc.) are simultaneously a substrate for different groups of microorganisms, so it can be argued that in the bioreactor space, as in the natural microbiological system, these groups of bacteria simultaneously consume these compounds and competition for them.

Many researchers define the limiting stage of bioconversion of organic matter (OM) as hydrolysis. During hydrolysis, cell wall destruction and degradation of extracellular polymeric substances occur, forming available organic materials for acid-forming microorganisms. This process is especially important for the decomposition of wastewater because the main component of its organic fraction is fibre—an unfavourable substrate for microbial degradation [37].

On the other hand, most species of methane-forming archaebacteria have lower growth rates than microorganisms that function in other stages of the anaerobic fermentation process [38]. Therefore, the effective conversion of intermediates (simple organic acids, hydrogen and acetate) to CH_4 can be considered a stage that limits the speed of the whole process for decomposing organ-containing waste, without considering their specifics (composition). Acetic acid also forms in different places where methane is formed. Acidification of the environment in wastewater treatment plants (this problem also occurs during the operation of methane tanks) is probably associated with the activity of bacteria that convert CO_2 and molecular hydrogen into acetic acid (acetogenic bacteria). Therefore, the definition of the limiting stage is associated with experimental studies of certain enzymatic processes.

The microbial system enters the dynamic equilibrium phase only in the presence of balanced interactions at the substrate level and metabolic products between different groups of microorganisms. For example, the presence of acetoclastic methanogens or sulfate reducers capable of assimilating acetate at low concentrations is critical for the efficient operation of the anaerobic group, which does not allow its accumulation in the reactor in the future. It should be noted that first of all, hydrogen is utilized in the system, and acetate begins to be utilized only with a big delay. Therefore, the relationship between hydrogenotrophic and acetoclastic methanogens is very important.

The biochemical model of anaerobic destruction of organic waste in Fig. 2.3 shows the substrate-product bonds that arose during the imposition of schemes of methanogenesis, homoacetogenesis and sulfidogenesis, which is caused by the content in organic waste of substances (sulfur compounds, etc.) consumed by different groups of microorganisms.

It should be noted that when assessing competition, it is necessary to consider the initial ratio of biomass of competing organisms. This is especially important

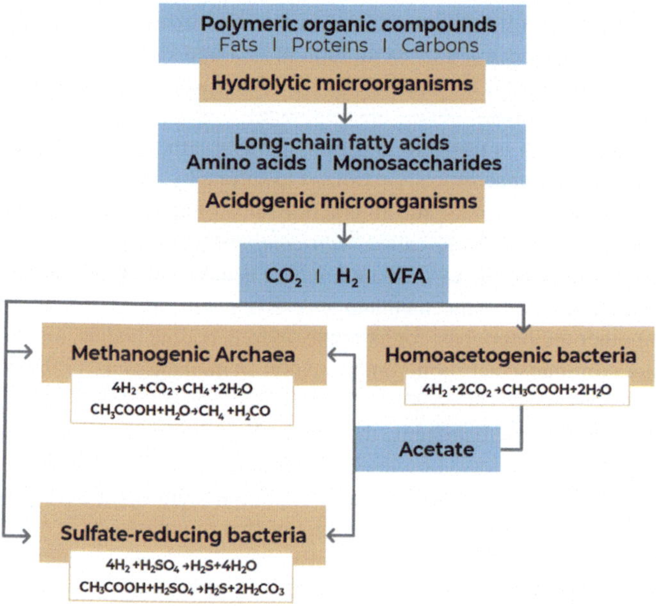

Fig. 2.3 Biochemical model of anaerobic grouping of destructive microorganisms

for slow-growing organisms. *Surplus organisms can be obtained even with these conditions' less favourable kinetic characteristics.*

An important place is occupied by sulfate-reducing bacteria (SRB) in the anaerobic biosorption of heavy metal (HM) ions. This is one of the mechanisms of SRB resistance to the action of HM ions due to their reduction into insoluble or sparingly soluble forms of sulfides using its metabolite-hydrogen sulfide.

Biogenic hydrogen sulfide in aqueous solution interacts as hydrosulfide/sulfide ions with HM ions. Chemical reactions occurring in sediments:

$$Me^{2+} + S^{2-} \rightarrow MeS \downarrow, \qquad (2.45)$$

where Me is the general designation of HM ions; MeS is the general designation of HM sulfides.

As a result of SRB activity, sulfides of cobalt, nickel, cadmium, iron, lead, zinc, and others are deposited from wastewater. This mechanism of resistance of bacterial cells to the action of toxic substances can be effectively used for precipitation and removal of HM from industrial wastewater.

Thus, if the group develops under sulfate conditions, the place of methanogens is occupied by sulfidogens. They yield H_2, ensuring the operation of syntrophs and a favourable thermodynamic balance of the group. *The process of sulfate reduction is more energetically advantageous compared to methanogenesis.* The result of the participation of sulfidogens is the formation of H_2S, which has an inhibitory effect on most groups, except for thiophilic, which oxidizes hydrogen sulfide to sulfate. For a methanogenic group, the usual concentration of 2–3 mm H_2S and its excess require changes in the group with the appearance of other species in the same functional groups of microorganisms.

Sulfate regeneration is possible in different ways: either H_2S is oxidized under aerobic conditions, or anaerobically oxidized to sulfate by anoxygenic phototrophic bacteria. The characteristic of the interaction of sulfidogenic and methanogenic groups is the existence of the so-called "*non-competitive C-pathway*". For example, pectin decomposition produces methanol, which is not used by sulfate reducers but is oxidized by methylotrophic methanogens. Even with an excess of sulfate and suppressing other methanogens, for example, in hypersaline bodies, methylotrophic methanogens continue to act "non-competitively" in methanogenesis and be a source of methane.

In contrast to SRB methanogens, they depend on the presence of an exogenous source of electron acceptors, i.e., the concentration of sulfate and the rate of its supply. In the absence of sulfate, SRB can function as syntrophic acetogens. In this case, they grow with methanogens as organisms that remove hydrogen.

Thus, in the anaerobic biotechnological system, substrate-product interactions arise during the superposition of trophic schemes of methanogenesis, homoacetogenesis, and sulfidogenesis. This should be taken into account in order to prevent emergencies in the system when predicting the process of bioconversion of organic waste to find appropriate levers to stimulate the development of necessary types of microorganisms, such as methanogens if biotechnology is aimed at biogas production, or sulfidogens if the removal of heavy metals is of primary importance.

2.2.5 Bacterial Oxidation Filter

When the products of anaerobic metabolism enter the zone where O_2 is present, they become available for oxidation by aerobes. This group of microorganisms is called "bacterial oxidation filter" due to the fact that an important part of the products of anaerobic metabolism is represented by reduced gases—"gasotrophs". Aerobic oxidation of metabolic products of secondary anaerobes is an alternative to their phototrophic oxidation. The group of gasotrophs does not form clear connections with each other, like anaerobes-organotrophs, because, in essence, it uses an allochthonous substance that enters its development zone from the outside. The common substrate for competition is O_3. Accordingly, gasotrophs are a series of gradations from aerobes to microaerophiles and facultative anaerobes. This series is expressed in the placement of gasotrophs in the form of films or thin zones in the area of oxycline or oxygen jump due to a sharp decrease in the concentration of O_2. Therefore, they are classified as so-called gradient organisms that develop in the transition zone from reduced to oxidative conditions. The most characteristic representatives of this group are colourless serobacteria, the layers of which are often coloured bright white with sulfur deposits. Gradient organisms are very difficult to cultivate in the laboratory.

Gasotrophs play the role of a protective filter against oxygen for the anaerobic group, preventing the penetration of O_2 into the lower layers. They are the cause of chemocline in stratified reservoirs, which is a geochemical barrier for mobile compounds. Oxidizable compounds are CH_4—for methanotrophs, H_2—for hydrogen bacteria, CO—for carboxydobacteria, NH_3—for nitrifiers, H_2S—for serobacteria resistant to oxidation of sulfur compounds, as thiosulfate—for thionic bacteria. Most gasotrophs are highly specialized organisms, but some, such as hydrogen bacteria, are facultatively autotrophic organisms. Gasotrophs are a critical group for the atmosphere's composition, preventing the release of gases such as methane [39].

As a result of the activity of thionic bacteria, as well as colourless and coloured serobacteria, a large part of hydrogen sulfide and other sulfur compounds are oxidized [40]. Moreover, in some cases, there is a significant amount of sulfur. Active sulfur oxidation by thione bacteria in the soil has gained practical application. For example, to reduce the alkalinity of the soil, elemental sulfur is applied, which is rapidly oxidized by these microorganisms with the formation of sulfuric acid. Studies on the use of bacteria of the genus Thiobacillus in reagent-free biocatalytic purification of hydrogen sulfide gases [41, 40].

2.2.6 Application of Microbial Associations in Biotechnologies of Environmental Protection

The basis of biotechnological methods of environmental protection is the variety of metabolic pathways of microorganisms: among the many options, there is at least one representative capable of disposing of the most unusual, including toxic compounds.

Recent decades have been characterized by the entry into the environment of a vast number of synthetic organic compounds (xenobiotics) of different structures, often toxic to living organisms. Some xenobiotics accumulate in the environment due to a certain lag in the evolutionary processes of microorganisms aimed at developing their ability to catabolism, i.e. the degradation of these compounds. Currently, the task is to maximize the potential of microorganisms that have a huge variety of enzymes that carry out the destruction of various xenobiotics. The use of this potential is "bioremediation".

If there are no microorganisms in the soil or water contaminated with xenobiotics capable of effectively degrading these compounds, it is advisable to introduce biodestructive microorganisms there. Microorganisms used for this purpose can be:

- natural, returning to the environment.
- selected.
- constructed using natural methods of horizontal transfer of genetic information (conjugation, transformation, transduction).
- obtained by molecular breeding.
- designed methods of genetic engineering.
- obtained using the method of protein engineering.

- obtained by a combination of these approaches and methods.

Although the vast majority of synthetic organic compounds have only recently appeared in the environment, microorganisms capable of degrading some of these compounds already exist in nature.

Biodegradation of oil pollution on soil and water

In case of emergency oil spills, the contaminated areas are treated with a specially grown association of oil-oxidizing microorganisms, making various additives for their nitrogen and phosphorus nutrition. This allows the utilization of petroleum carbohydrates, converting them into biomass of microorganisms and carbon dioxide.

As we can see, the most common is pollution with oil and various petrochemical products [42, 43]. There are a number of technologies for the treatment of contaminated soils. The most common are incineration, disposal and biooxidation. The available comparative assessment shows that biooxidation is the cheapest, most environmentally friendly, promising method. The literature uses the terms "biodegradation", "bioregeneration", "biore reclamation", and "bioremediation" to denote this process.

Biological wastewater treatment

There are microorganisms for which the contaminants contained in wastewater are nutrients. A method of aerobic biological wastewater treatment using activated sludge—a group of microorganisms has been developed [44]. Large volumes of domestic and industrial effluents are processed in large tanks with mixing and constant aeration of the liquid. Conventional cleaning removes mostly organic contaminants. Additional treatment measures are required if the effluents contain heavy metals (copper, nickel, cadmium, chromium, lead, etc.). There are certain types of microorganisms on which metals dissolved in a liquid can settle (sorb). The concentration of metals increases so much that the biosorbent can be considered a raw material for non-ferrous metals after heat treatment.

The above-described introductory provisions of the equations of enzymatic catalysis can be applied to the formation of ideas about the mechanism of complex biochemical transformations of industrial wastewater under the influence of activated sludge microorganisms. It should be borne in mind that the biochemical process of oxidation of pollutants in the aeration tank is a complex of different reactions that occur sequentially with the participation of a large number of different in nature enzyme systems.

In wastewater treatment technology, the substrate is understood as the concentration of pollutants in the treated wastewater, and the concentration of activated sludge is equivalent to the concentration of enzymes [45]. The idea of the kinetics of oxidation of organic matter in industrial wastewater is complicated by the conditions of biochemical reactions, i.e. hydrodynamic conditions of the treatment process due to design features and used aeration tank (mixer, displacer), as well as a number of other reasons. Therefore, the calculation of the aeration tank must be performed at the rate of the total enzymatic reaction. However, to develop a scientifically sound regime of biochemical oxidation of wastewater from chemical plants of multicomponent composition, it is necessary to identify kinetic characteristics that describe the dependence of the specific oxidation rate on the value of the substrate. The obtained ideas about the maximum rate of oxidation of the substrate allow you to choose the optimal mode of operation of aeration tanks during the treatment of the studied wastewater.

The Urrea et al. [46] research has shown that the kinetics of oxidation of chemical wastewater production by activated sludge is often described by a sigmoid curve (wastewater produced by caprolactam, formaldehyde, etc.). This indicates that the active enzyme centers have low catalytic activity at low substrate concentrations. Analysis of the oxidation kinetics of individual components of wastewater allows you to choose the method and scheme of treatment, determine the technological mode of operation of biochemical treatment plants, address the need to remove from the mixture of wastewater most toxic components and choose the method of local treatment.

Differentiated kinetic analysis of a mixture of three types of wastewater from a synthetic rubber plant allowed us to obtain numerical values of kinetic constants for all categories of investigated wastewater and, based on these values, to determine the process parameters and operating conditions of treatment plants.

The presence of nitrogen in industrial wastewater stimulates the metabolic activity of nitrifying bacteria in activated sludge, which, under aerobic conditions, oxidize ammonium nitrogen to nitrates. Stoichiometrically, these reactions require 4.57 g of oxygen per 1 g of oxidized ammonia. The nitrification process in wastewater treatment plays an important role because it is the final stage of oxidation processes occurring in the aeration tank [47]. Nitrogen is needed by microorganisms in large quantities, as its content in bacteria is approximately 10 g per dry biomass.

The assimilation of ammonia and nitrates by pure cultures of microorganisms is subject to the Michaelis–Menten kinetic equation [48]. In an environment with a high concentration of ammonia, microorganisms consume it due to reductive amination of the intermediate product of the cycle of tricarboxylic acids, aoxoglutarate. The enzyme glutamate dehydrogenase catalyzes the reaction. It was found that at low concentrations of ammonia, this enzyme does not participate in the primary assimilation of ammonia because it is inactive at low concentrations of the latter. At low concentrations, ammonia is assimilated by another enzyme, glutamine synthetase.

Nitrogen occurs in industrial and domestic wastewater in the form of ammonia, nitrite, nitrates, and urea, as well as in several organic nitrogen-containing compounds of natural and unnatural origin. The preferred nitrogen source for the heterotrophic population of activated sludge is ammonia; oxidized forms of nitrogen are first subjected to the processes of reduction to ammonia and then involved in biosynthesis. Organic compounds are pre-cleaved with the release of ammonia, which is then used in the consumption process.

Nitrate reduction can occur in wastewater treatment only under anaerobic conditions or at low oxygen concentrations, i.e. oxygen inhibits the activity of enzymes that carry out dissimilation reduction of nitrates. Denitrification is characterized by the complete oxidation of the organic substrate to CO_2. The appearance of denitrification in wastewater treatment in the aeration tank indicates an imbalance in the ratio of organic matter and dissolved oxygen. The latter causes local oxygen-free zones in the aeration tank and the secondary settling tank, which are active parts of denitrification processes. Denitrification has a negative effect on the operation of secondary settling tanks, as the consequences of this process are the emergence of large masses of activated sludge permeated with gas bubbles and the removal of sludge into the reservoir. At the same time, in the practice of wastewater treatment, the denitrification process is widely used as a local method of wastewater treatment and their purification from excess nitrogen compounds before their release into the reservoir [49].

Sulfur is a necessary mineral element for the life of microorganisms. It is part of proteins and several other substances important for the physiology of bacterial nutrition. A number of categories of chemical wastewater contain oxidized sulfur compounds in the form of sulfates [50]. Heterotrophic activated sludge microorganisms can metabolize sulfates with the subsequent incorporation of sulfur into the cellular material. In the conditions of insufficient access to oxygen and long stay of return sludge in secondary settlers and receiving chambers of pump stations in the presence of sulfates, sulfate-reducing bacteria which convert sulfates on H_2S develop [51]. Specialized anaerobes and a number of common bacteria form this group of microorganisms.

In the presence of iron, hydrogen sulfide reacts with the latter, forming sulfides, which can colour the activated sludge blue or black. Similar phenomena are observed in the operation of chemical plants' treatment facilities. They can be evidence of a violation of the technological regime at the stage of settling and returning activated sludge.

Biocomposting of solid organic waste

An analogue of aerobic wastewater treatment is the aerobic biocomposting of solid waste. Solid waste can be inoculated from a mixture of groups of different types of destructive microorganisms' pollutants of different nature, as well as contribute ballast material such as peat. This allows you to turn waste into fertilizer or use it to fill roads, construction sites, and other cases.

The basis of biothermal treatment is the self-heating of solid waste (solid waste) to 50 °C and above under the influence of thermophilic (heat-loving) microorganisms [52]. This occurs in loose organic waste content with a humidity of 40–45 to 70–75% with difficult heat transfer. Usually, such conditions are created by themselves when storing large amounts of wet hay, straw, leaves, tops, peat, horse manure, vegetables, potatoes, household waste, etc. The exact process has long been used to heat greenhouses and hothouses with horse manure for the early distillation of vegetables.

Composting can significantly reduce fuel and energy costs for decontamination of precipitation and improve their sanitary and hygienic performance (due to the death of pathogens, helminth eggs and fly larvae). In aerobic microorganisms' life process, organic matter is used and consumed, so the biothermal process is most effective during composting of raw unfermented precipitation. It is possible to use the process of biothermal treatment in combination with anaerobic fermentation of precipitation in mesophilic conditions.

The composting process consists of two phases. The first phase lasts 1–3 weeks and is accompanied by intensive development of microorganisms, and the sediment temperature rises to 50–80 °C. At the same time, a deposit is disinfected and its weight reduced. The second phase—maturation of compost—is longer. It lasts from two weeks to 3–6 months and is accompanied by the development of protozoan and arthropods, lowering the temperature to 40 °C and below. Increasing the ambient temperature intensifies the decomposition of organic matter [53].

For uniform heating and provision of microorganisms with air during the composting period, 2–3 times mixing of the composted mass is required. Depending on the composition of the sludge, duration and conditions of composting, the amount of organic matter is reduced by 25–40%.

As a result of the biothermal treatment process, compost is obtained in the form of bulk material with a moisture content of 40–50%. The finished compost has no smell, does not rot and is a good fertilizer.

Organic wastes with a large amount of water, such as cow manure, pig manure, raw sewage sludge with a moisture content of more than 75%, etc., are poorly heated and do not give high temperatures because they lack air. For biothermal processing, there is a need to add dry material, such as leaves, peat, straw, dry waste, etc. Dry waste, on the other hand, needs to be moisturized. The most intensive biothermal processes occur in a weakly alkaline reaction, so adding slaked lime to the waste at approximately 1 kg per 1 m^3 of waste is recommended. At temperatures above 50 °C, humification processes take place intensively; nitrification processes are not used here, so the resulting biothermal waste processing and subsequent fertiliser screening are very valuable for agriculture humus.

Biothermal methods of solid waste disposal include [54]:

- processing of solid waste without its preliminary preparation by composting in stacks, in a tubeless way with aeration and in biothermal chambers, and also in special installations.
- processing of solid waste with its preliminary preparation and subsequent disposal in stacks or in special installations.

Biogeotechnology of metal leaching

Associations of thionic (oxidizing sulfur and sulfur-containing compounds) bacteria are mainly used to extract metals from ores, ore concentrates and rocks. During the processing of poor and complex ores, thousands and even millions of tons of precious metals are lost in the form of waste, slag, and “tails”. There are also emissions of harmful gases into the atmosphere. Bacterial-chemical leaching of metals reduces these losses. The basis of this process is the oxidation of sulfide minerals contained in ores by thione bacteria [55]. Oxidized sulfides of copper, iron, zinc, tin, cadmium, etc. In this case, metals from insoluble sulfide form are converted into sulfates, which are well soluble in water. Metals are extracted from sulfate solutions by precipitation, extraction, and sorption. One of the possible ways to extract metals from solutions is the adsorption of metals by cells of living microorganisms, the so-called biosorption of metals [56, 57]. Metals are part of specific proteins—metallothioneins.

2.3 Viruses and Their Use in Greening Processes of Agricultural Greening

Baculoviruses are double-stranded DNA viruses; in three groups, there are biopesticides: nuclear polyhedrosis viruses (NPV), granulosa viruses (GV), and filter viruses [58]. The production of viral drugs is based on the mass reproduction of the host insect on artificial media. There are two methods of using viral drugs: introduction of viruses into dense populations of insects in relatively small areas and treatment of infected areas by spraying or pollination in the early stages of larval development. Improvements and development of insect cell culture technology, as well as the selection and even creation of new viruses, given the production of eukaryotic viruses in prokaryotes, can affect the competitiveness of viral pesticides compared to chemicals.

2.4 Enzymes

2.4.1 Obtaining and Using Enzyme Systems in Environmental Biotechnology

A special group of biological agents in biotechnology—enzymes, so-called catalysts of biological origin. Enzymes and enzyme systems are widely used in various industries, such as medicine, agriculture, chemical analysis, etc.

Microbial enzymes are obtained by growing microbial cells under special conditions so that the cells produce the maximum amount of active enzymes. To maintain the integrity of a large number of enzymes, it is important to monitor environmental conditions during production.

After stopping the growth of bacterial cells and their production of enzymes, they are separated by specially designed methods. Enzymes are further purified to the desired state.

Waste treatment levels usually involve any environmental conditions under which enzymes should function normally. For optimal use of the catalytic potential of enzymes, it is necessary to get acquainted with the basic principles that affect their activity and stability.

Many environmental factors affect the activity of biological enzymes produced for commercial use [59]. The pH of the environment has the strongest effect on the activity and stability of enzymes, and its optimal level for different enzymes varies greatly. For enzymes produced by most microorganisms used for commercial purposes, the optimum is in the range of 4.0–7.5 pH. Another important factor influencing the activity and stability of enzymes is temperature. Because enzymes are biochemical catalysts consisting mainly of protein, they are sensitive to temperature. Increasing the ambient temperature to the optimum increases the degree of enzyme activity. However, higher temperatures lead to rapid enzyme degradation with a subsequent irreversible decline in activity.

Enzymes are substances of a protein nature and are, therefore, unstable during storage and sensitive to thermal effects. In addition, enzymes cannot be used often due to difficulties in separating them from the reagents and reaction products. The creation of immobilized enzymes helps to solve these problems. The beginning of this method dates back to 1916 when J. Nelson and E. Griffin absorbed invertase carbon and showed that it retains catalytic activity in this form. The term "immobilized enzymes" was legalized in 1971, and it means any restriction on the mobility or movement of protein molecules in space.

The essence of enzyme immobilization is their attachment in active form to the insoluble base or placement in a semi-permeable membrane system. The enzyme is attached to the carrier by adsorption, chemical bonding or mechanical incorporation of the enzyme into an organic or inorganic gel (in a capsule, etc.) [60]. The attachment of the enzyme is allowed only due to the functional groups that are not part of the active center of the enzyme and do not participate in the formation of the enzyme–substrate complex. The enzyme carrier or matrix may be in the form of granular material, fibrous structure, lamellar surface, films or tissues, hollow fibers, tubes, capsules, etc. The size of the carrier particles is important. It is important to have a large surface, so small particles with a 0.1–0.2 mm diameter are recommended. The enzyme carrier can be both a natural substance and a synthetic polymer.

Advantages of immobilized enzymes [61, 62]:

1. A heterogeneous catalyst is easily separated from the reaction medium, which makes it possible to stop the reaction at any time, reuse the enzyme, and obtain a pure product from the enzyme.
2. The enzymatic process using immobilized enzymes can be performed continuously by adjusting the catalysed reaction rate and the product yield.
3. Modification of the enzyme purposefully changes its properties, such as specificity (especially relative to the macromolecular substrate), the dependence of catalytic activity on pH, ionic composition and other parameters of the environment, and stability to denaturing actions.
4. Regulation of the catalytic activity of immobilized enzymes by changing the properties of the carrier by the action of physical factors such as light and sound. Immobilization of enzymes is carried out by binding on insoluble carriers by intramolecular or intermolecular crosslinking of protein molecules by low molecular weight bifunctional compounds and by attachment to a soluble polymer.

Enzymes are increasingly used in various biotechnological processes and industries, but by the 60 s of the twentieth century, this direction was constrained by difficulties of their reception, instability, and high cost. As a separate field in creating and using new biological agents, it is necessary to identify immobilized enzymes, which are harmoniously functioning systems. The correct choice of enzyme, carrier, and immobilisation method determines the action of these enzymes. The advantages of immobilized enzymes over soluble ones are stability and increased activity, retention in the reactor volume, the possibility of complete and fast separation of target products and organization of continuous fermentation processes with repeated use of biological agents.

Immobilized enzymes open up new possibilities for creating biological microdevices for use in analytics, energy conversion, and bioelectrocatalysis.

For degradation and modification of anthropogenic organic compounds entering the environment, enzymes of different classes are used, in particular: lipase—which decomposes fats and oils; protease—which decomposes proteins; cellulase—which decomposes cellulose; amylase—decomposes carbohydrates and starch as well as ligninase, tyrosinase, monooxygenase, dioxygenase and others [63]. For a deeper understanding of the mechanisms of enzyme systems and the possibility of their use, consider some kinetic features of enzymatic reactions.

2.4.2 Biophysical Bases of Enzymatic Reactions

The main way to increase the reaction rate is the use of a temperature factor or catalysts. In the case of inorganic or synthetic catalysts, it is necessary to use mostly high temperatures. However, biocatalysts work effectively at relatively low temperatures. Reaction rates depend on the concentrations of the substrates. When the enzyme is saturated with the substrate, it cannot function faster—this is the maximum reaction rate (V_{max}), i.e., if the proportion of free enzyme is extremely low.

The total rate of the enzymatic reaction should be proportional to the concentration of the enzyme–substrate complex *ES* (*E*—enzyme, *S*—substrate) [64].

The dependence of the rate of the enzymatic reaction on the concentration of the substrate is graphically represented by a hyperbolic curve (Fig. 2.4). K_m in the figure is the Michaelis–Menten constant, i.e. the concentration of a specific substrate at which a particular enzyme provides a reaction rate equal to half V_{max}. Until now, the analysis of the kinetics of enzymatic reactions is carried out based on the following mathematical equation Michaelis—Menten:

$$V_o = \frac{V[S]_{max}}{K_m + [S]} \tag{2.46}$$

where V_0—initial velocity at concentration S; V_{max}—maximum speed; K_m—Michaelis–Menten constant for this enzyme, corresponding to a certain substrate. This equation, in its form, corresponds to the Monod equation.

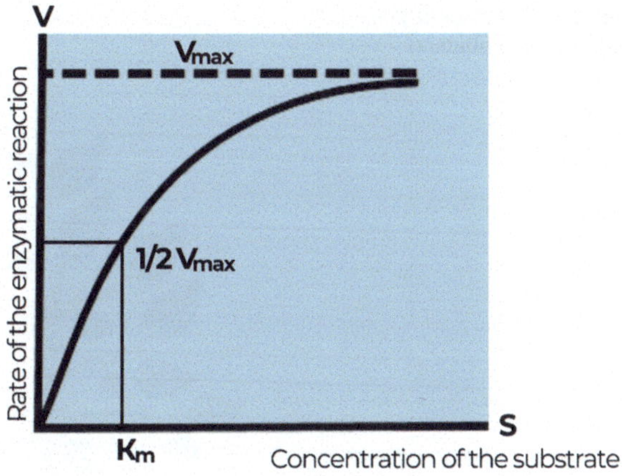

Fig. 2.4 Dependence of the rate of the enzymatic reaction on the concentration of the substrate

The Michaelis–Menten equation is often transformed to obtain advantages in studying kinetic enzymatic equations. Such transformations include the Lineweaver–Burk equation:

$$\frac{1}{V} = \frac{K_m}{V_{max}\frac{1}{S}\frac{1}{V_{max}}} \tag{2.47}$$

According to this equation on the graph constructed in double inverted coordinates, it is possible to define V_{max} *more precisely* (Figs. 2.4 and 2.5).

Thus, the range of processes in which enzymes or enzyme systems can be used is very wide. The development of environmental biotechnology is impossible to imagine without biocatalytic reactions of different levels, associated primarily with the biodegradation of various pollutants and the implementation of low-waste industries [65].

2.5 Plants

2.5.1 Phytoremediation Processes

Phytoremediation is a technology for the restoration of a polluted environment using various types of plants [63]. Phytoremediation technology is used directly in the area of pollution, contributing to cost reduction and minimizing the contact of contaminated xenobiotics with humans and the environment. The most important thing is that the soil does not lose its fertility after phytoremediation. Therefore, this technology is environmentally friendly and cost-effective [66].

The plant affects the environment in different ways [67, 68]. The main ones are (Fig. 2.5):

Fig. 2.5 Processes of plant influence on pollutants

- rhizofiltration—the roots absorb water and chemical elements necessary for plant life.
- phytoextraction—the accumulation of dangerous contaminants in the plant's body (e.g., heavy metals).
- phytovolatilization—plant leaves' evaporation of water and volatile chemical elements (As, Se).
- phytotransformation:
- phytostabilization—the conversion of chemical compounds into less mobile and active forms (reduces the risk of contamination).
- phytodegradation—degradation by plants and symbiotic microorganisms of the organic part of pollutants.
- phytostimulation—stimulation of the development of symbiotic microorganisms involved in the purification process.

Microorganisms play an important role in the degradation of contaminants. The plant serves as a biofilter, creating a habitat for them (providing access to oxygen and loosening the soil). In this regard, the purification process also occurs outside the growing season (in the non-summer period) with slightly reduced activity.

Currently, there is active research on hyperaccumulators. One of the most promising is technical hemp [69].

The reasons why the plant is superior to other phytoremediators are:

- *Cannabis sativa* (hemp) grows rapidly, reaching full maturity in 100–180 days.
- develop a broad root system that extends 1.5–2 m into the ground. Hemp completely removes toxic substances at this depth without needing additional soil reclamation.
- the ability of hemp to grow regardless of the presence of toxins in the soil.
- ability to bind hazardous substances in the air and soil. Hemp removes carbon dioxide from the air just as a plant absorbs heavy metals and other soil contaminants.

Advantages of phytoremediation in comparison with traditional methods:

- possibility of in situ remediation.
- relatively low cost of work performed, compared to traditional treatment plants.
- method safe for the environment.
- theoretical possibility of extraction of valuable substances from the green mass of plants (Ni, Au, Cu).
- the ability to monitor the cleaning process.
- the level of treatment is not inferior to traditional methods, especially with a small amount of wastewater (for example, in villages).

One of the key points of phytoremediation is the optimal composition of tolerant plant species that can survive in the conditions of pollution and transform and neutralize them. The choice of plants for this technology is determined by their ability to bring groundwater to the surface due to evapotranspiration to break down contaminants with their enzymes and accumulate these compounds in biomass.

The effectiveness of soil phytoremediation depends on plant productivity. With more biomass, more pollutants that enter the plant are removed from the soil. The biomass of *Brassica juncea L.* plants increased with increasing concentrations of added $CoSO_4 \cdot 7H_2O$ and $NiSO_4 \cdot 7H_2O$. growth retardation is more noticeable; at high concentrations (15 threshold limit value, (TLV)), the plants are suppressed and do not reach the required biomass.

Recently, phytoremediation has received considerable attention and recognition. Its main advantages include economic feasibility, efficiency, and environmental sustainability. Although phytoremediation possesses several benefits compared to other remediation technologies, it also has some limitations. Phytoremediation is a slow process that relies on biological cycles and requires several months to years and multiple agricultural crop cycles for soil recovery [70]. This approach suits less or moderately contaminated areas where plant toxicity remains relatively low, and vegetation can thrive.

The change in root mass according to the variants during cobalt contamination also occurred differently, depending on the dose of the contaminant. In the plants of *Brassica juncea L.*, like the mass of the aboveground part, there was a significant increase (16% compared to control, $P \leq 0.05$) root mass at an average level of pollution, with increasing pollution there was a decrease, especially this adverse effect (88% compared with control) was expressed when making cobalt in the soil in doses of 10 and 15 TLV. Similar results were observed during nickel contamination of the soil.

The most important components of the technology of restoration of soil contaminated with organochlorine pesticides with the help of plants are phytoextraction and phytostabilization. The plant organism's phytoextraction potential depends on the pollutant's hydrophobicity. The degree of hydrophobicity largely determines the efficiency of absorption and movement of the pollutant in plants. Persistent organic pesticides, such as 1,1,1-trichloro-2,2-bis(p-chlorophenyl)ethane (DDT) and hexachlorocyclohexane (HCH), are classified as toxicants. In the soil, they bind to organic or inorganic compounds (chelate) and become isolated within the soil's natural solid particles, reducing the bioavailability of flora.

To increase the efficiency of phytorementation, it is proposed to use substances that stimulate plant growth, increase the mobility of hydrophobic contaminants and increase the rate of their entry into plants [71]. In this regard, heterocyclic compounds of the oxane series that potentially have biological activity have been synthesized.

The process of absorption by certain plants of organic residues and other pollutants contained in wastewater is characteristic of natural biosystems. The most famous plants that purify water in natural conditions in water bodies (swamps, ponds and lakes) include duckweed, water spinach, wolfia, and rootless duckweed. During wastewater treatment, the following species of higher aquatic plants are most often used: bulrush, lake reeds, narrow-leaved and broad-leaved cattail, curled and comb-leaved pondweed, greater duckweed, elodea, yellow iris, flowering rush, arrowhead, amphibious bistort, hornwort, chara, iris, and others [72].

There is a method of phytoremediation for the removal of metals from the soil due to the accumulation of metals by plants, which is an alternative to existing methods of soil restoration (excavation, washing), in particular, due to low costs. This is especially important for mountain-depleted mountain areas, as it improves the landscape.

Different species of higher aquatic vegetation enrich the water with oxygen obtained through biosynthesis, and toxic impurities are broken down into constituent chemical elements. Thus, wastewater is detoxified. With the help of higher aquatic vegetation, this purification method is implemented in bioplateau systems [73].

Further research and development are essential to overcome the limitations of phytoremediation and make it more effective and efficient. Nevertheless, the potential benefits of phytoremediation are substantial. This is a natural, cost-effective, and environmentally friendly method for cleaning contaminated sites. It can remove various pollutants, including heavy metals, organic contaminants, and radioactive contamination. Moreover, it has the potential for large-scale application in solving the global problem of environmental pollution [74].

2.5.2 Bioplateau Technology

Bioplato is an engineering structure used for the treatment and post-treatment of domestic and industrial wastewater and contaminated surface runoff, which does not require (or almost does not require) electricity consumption or the use of chemical reagents in low maintenance.

The technology's basis is the natural self-purification processes inherent in aquatic and near-water systems. The principle of "bioplateau" technology is to use higher aquatic vegetation [75]. During wastewater treatment, such types of higher aquatic vegetation as a bulrush, lake reed, narrow-leaved and broad-leaved cattail, comb-leaved and curled pondweed, greater duckweed, elodea, water hyacinth (Eichhornia), iris, etc. are used. Higher aquatic plants (HAP), such as bulrush, reeds, cattail, have the ability to remove contaminants from water: nutrients (nitrogen, phosphorus, potassium, calcium, magnesium, manganese, sulfur), heavy metals (cadmium, copper, lead, zinc), phenols, sulfates, petroleum products, synthetic surfactants, and improve indicators of organic pollution such as biological oxygen demand (BOD) and chemical oxygen demand (COD). For example, the cattail root system has a high accumulative capacity for heavy metals. The concentration of metals in the root system of cattail, which grew on the shores of sludge storage power plants, reached (mg/kg): iron—199.1,manganese—159.5; copper—3.4; zinc—16.6. It is known that reeds have high adaptive properties and can germinate in very contaminated industrial wastewater areas. It removes a number of organic compounds from water, including phenols, naphthols, anilines and other organic substances. The specific absorption of minerals by reeds reaches (1 g per 1 g of dry weight): calcium—3.95; potassium—10.3; sodium—6.3; silicon—12.6; zinc—50; manganese—1200; boron—14.6.

Bioplato with HAP have a significant oxidizing ability due to the creation of a biofilm of aquatic organisms (periphyton) on the surface of the inert substrate and the immersed part of the rhizomes and stems of HAP, which interact symbiotically. Part of the biocenosis of microorganisms is in a suspended state in the form of flakes and also forms a layer of natural sediments—benthos, in which there is an active process of anaerobic decomposition of organic pollutants. A significant role in the purification process is played by saprophytic bacteria, which, together with HAP, successfully perform the function of disinfectants due to their metabolic products and antagonism with heterotrophic bacteria, which in some cases avoids the use of chlorination or ozonation of water. Bioplato wastewater treatment systems usually consist of several blocks arranged in a cascade, and the surface bioplate block is finite. The bioplato's final structure can include a swampy area (natural surface bioplateau) with sufficient thickets of higher aquatic vegetation. The initial block of structures is a settling tank, where large inclusions and suspended solids are removed. The factors that most affect the efficiency of treatment include water and air temperature, pH and Eh of the environment, time of year, hydraulic load on structures, aeration, and the initial concentration of pollutants in the water supplied for treatment. Purified water flows into the reservoir. The quality of reservoirs after treatment in almost all standardized indicators meets the regulatory requirements for the communal category of water use.

Consider a few examples of successful use of plants for wastewater treatment.

In Norway, 40 km south of Oslo, an experimental bioplato was built to treat agricultural surface runoff, based on a filter constructed of 8 parallel strips (each measuring 3 × 40 m) with a depth of 0.5 m and an area of 1200 m^2. The catchment area is 0.8 km^2. Previous studies have shown significant efficiency in removing suspended solids—45 to 75%, phosphorus—21 to 44%, and nitrogen—15%. Today, research is ongoing.

It is also important to note that vegetation can be considered a kind of test object when assessing the pollution levels of environmental components, in particular, atmospheric air, groundwater and surface aquifers in areas of open pit extraction of non-metallic minerals. The stability of biocenosis largely depends on living matter, the ability of biota to withstand the effects of external factors, its vulnerability and the possibility of recovery under certain conditions. That is why the indicator of changes in the state of biocenosis is phytocoenosis, which provides energy fixation and the cycle of substances. It is sensitive to changes in external factors and well reflects these changes visually.

In Ukraine, the use of HAP on different bioplato systems—which do not require (or almost do not require) electricity consumption and the use of chemical reagents with minimal periodic maintenance—began in the last century. Hydroplant systems with higher aquatic plants are widely used in the country, with the most common being engineering-biological systems based on a closed hydroponic bioplato. Experts from a scientific research center, together with representatives from an institute of hydrobiology, have developed various types of these systems, which have found broad application in different industries for wastewater treatment and post-treatment. Moreover, HAP systems, tested by specialists from an institute of hygiene and ecology, have been recognized as ensuring that the reclaimed water meets the standard quality required for domestic and aquaculture water bodies. Moreover, since 1997, a complex of bioplato constructions has been operating in a village (in the Kharkiv region), where the technology is used for wastewater treatment, achieving a productivity of 40 m^3 per day [76].

In addition to water treatment, bioplato, like other ecosystems, performs a number of ecological functions—hydrolytic, biological, biogeochemical. Being highly productive systems, they contribute to the maintenance of landscape and ecosystem diversity, the preservation of the abundance of life forms, the creation of habitats for various species of flora and fauna.

Further research is necessary to systematically study the treatment of surface wastewater by plants under the influence of various external factors. In order to reliably design treatment plants incorporating HAP systems, studies should be conducted to investigate not only the dynamics of purification but also the effects of different chemical compounds, ambient temperature, light, and other environmental parameters. This research should aim to create optimal living conditions for the plants through suitable engineering solutions, considering both natural factors (changes in temperature, light, etc.) and the operational conditions of treatment plants (changes in effluent flow, concentrations, etc.).

It should be noted that in biological systems, there are important symbiotic relationships between different groups of organisms, such as prokaryotes and eukaryotes. These relationships in the environmental biotechnology of the future will open various opportunities for creating sustainable self-regulating biological systems within the framework of environmental technologies.

2.6 Ecological Aspects of Microalgae Biocultures

In the conditions of intensification of agricultural production and sharp increase of anthropogenic influence on the environment, particularly on the soil cover, the role of biological factors in the increase of soil fertility and their reclamation considerably grows. The effective use and regulation of soil biota development, with algae as a stable and essential component, is highly beneficial [77].

Microalgae are successfully used to increase soil fertility, i.e. to replenish organic matter (such as humic acids), which helps to increase crop yields. Green and blue-green microalgae are used for this purpose. Using microalgae as biofertilizers is cost-effective and safer for the environment than chemical fertilizers. Unlike conventional chemical fertilizers, microalgae serve as a source of organic carbon (C) when introduced into the soil. This becomes an increasingly important aspect, considering that the depletion of organic C in the soil is a primary type of degradation of arable land, leading to decreased soil quality and fertility. During photosynthesis, microalgae absorb organic carbon, and their exopolysaccharides, released by many strains, act as a source and reserve of carbon, contributing to soil stabilization and aggregation [78].

Soil algae can also be an indicator of soil condition. Among them, cyanobacteria play a particularly crucial role due to their unique ability to enhance soil structure, boost nutrient availability, support plant health, and optimize the physicochemical properties of the soil (Fig. 2.6). These impacts, highlighting the multifunctional contributions of cyanobacteria to soil health and fertility, including forming aggregates, mobilizing nutrients, and improving soil resilience. Numerous data suggest that algae serve as test objects in determining the need for fertilizers in soil and in testing various pesticides, for example, to assess the residual toxicity of herbicides in the soil. Microalgae can be used to study the mechanism of action of various herbicides and model objects in the study of phytotoxic chemical compounds and the identification of their structure. They play a vital role in biological water treatment. Considering the economic efficiency, the most promising is the use of algae for wastewater treatment of food industry enterprises, fish farms, livestock farms, and poultry farms. As phototrophic organisms, they enrich the aquatic environment with oxygen, thereby helping to accelerate oxidative processes and mineralization of organic impurities in wastewater. One of the most effective ways to purify water is bioflocculation.

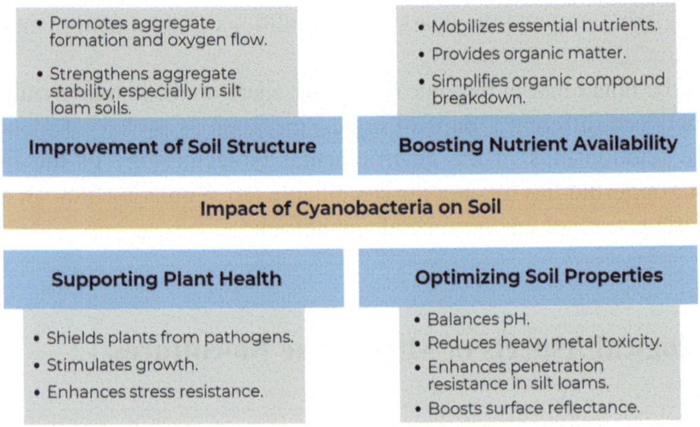

Fig. 2.6 Impacts of cyanobacteria on soil

One major limitation hindering the widespread adoption of microalgae in agriculture is the inadequate understanding of interactions between microalgae (and their extracts), plants, and the surrounding environment. Despite numerous studies conducted in this field, the vast diversity of photosynthetic organisms combined with a large number of metabolites that can be extracted from them presents a significant challenge to gaining a deep understanding of the impact and mechanisms of microalgae on soil and plants [79].

The key aspect of the development is to use a specially grown bioculture to rehabilitate aquatic ecosystems. In contrast to the known methods, microalgae in an inactive state are used because the physicochemical properties of biomass are used as a flocculant.

Algae is one of the richest sources of biofuel production. The oil yield from algae is about 50%, significantly higher than in rapeseed. The advantages of obtaining biodiesel from algae—high growth rate, resulting in high yield per 1 ha; algae-derived biofuels are also sulfur-free, non-toxic, and biodegradable.

Microalgae and cyanobacteria effectively enhance the content of nutrients, structure, and fertility of soils. Several microalgae strains have been selected based on characteristics such as CO_2 fixation and nitrogen and phosphorus cycling, which are beneficial to plants. However, their potential for improving arid lands and saline and sodic soils has not yet received widespread recognition. Similarly, their biological control properties, potential for improving crop yields and its components, and biosorption of organic carbon in the soil require additional research and technological innovations. The balance between benefits and risks associated with microalgae applications must be considered based on their genomic potential, particularly during prolonged use [80]. Nevertheless, a new hope for utilizing these resources of microalgae in agriculture lies in their unexploited surplus, which is underestimated by modern cultivation-based methods and a better understanding of their evolutionary advantages and ecosystem services. Future work should improve their collections of pure cultures, understand their metabolic lifestyle, study their environmental role, recognize their evolutionary significance, disseminate new knowledge, and raise awareness among farming communities and policymakers.

2.7 Sanitary and Hygienic Assessment of a Biological Object

2.7.1 Comprehensive Evaluation of Industrial Strains

The study of biological agent strains involves the study of their microbiological, technological, sanitary and environmental properties. The scheme of tests of the strains perspective for introduction in manufacture is given in Fig. 2.7. At the stage of microbiological research (stage 1), determine the taxonomic position of the strain, develop methods for its identification in the industrial process and the environment, determine the conditions of its dominance in the case of unprotected cultivation conditions.

The study of the technological properties of the strain (stage 2) is aimed at determining its industrially valuable properties. During sanitary and hygienic research (stage 3), the pathogenicity of a strain, its virulence, and the threshold limit value of aerosols of live and inactivated cells in workplace air and atmospheric air are assessed. Ecological toxicological research (stage 4) involves determining the viability of

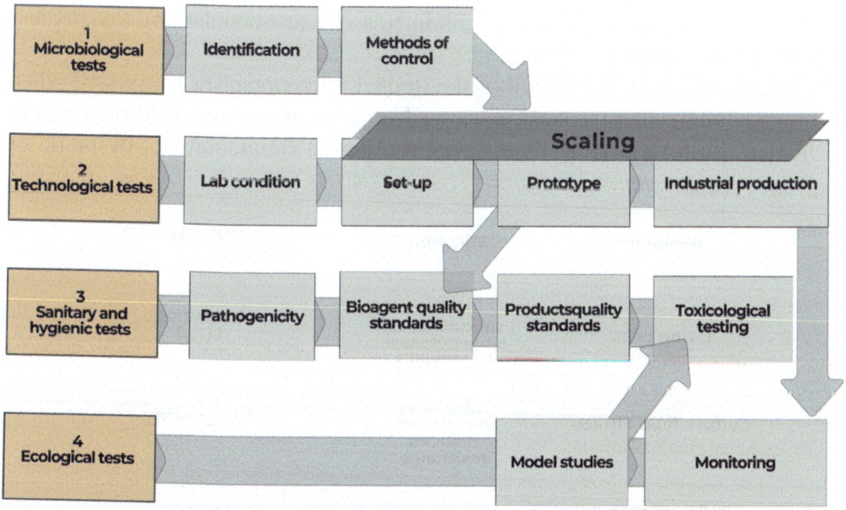

Fig. 2.7 Scheme of tests of new strains, promising for introduction into production

biological agent cells in ecosystems, studying their impact on the biocenosis of ecosystems—and determining the maximum allowable ecological load.

Hygienic standardization of microorganisms-producers and finished forms of preparations that contain them in the objects of production and environment differs significantly from the justification of sanitary standards of chemicals, including products of microbiological synthesis. Determining the Threshold Limit Values of chemical substances is not fully applied to microbial contamination due to the fundamental differences between chemical and microbial contamination.

The principles of determining the TLV of chemical substances are based on the principle of threshold [81]. The concept of threshold for many biological agents is absent. There are difficulties in quantifying the risk of the presence of a biological factor in the environment, as the entry into the susceptible organism of even a few living microbial cells of a pathogenic microorganism (for example, during strain identification or vaccine-related work) may be sufficient to cause any disease clinical picture and result.

2.7.2 *Substantiation of Threshold Limit Value of Living Cells of Microorganisms in the Air of the Work Zone and Atmospheric Air*

The research on the substantiation of the TLV of microorganism producers in the air of the work zone and the atmosphere involves several stages. In the first stage, signs of pathogenicity of the object are detected, virulence is determined by the value of LD_{50} (lethal dose required to cause death in 50% of a test population), toxicity and toxigenicity (Fig. 2.8).

In the second stage, the hazard of the studied microorganism is assessed when laboratory animals enter the body in ways adequate to the actual conditions of work and human life to select the limiting harm criterion (immunotoxic, dysbiotic or

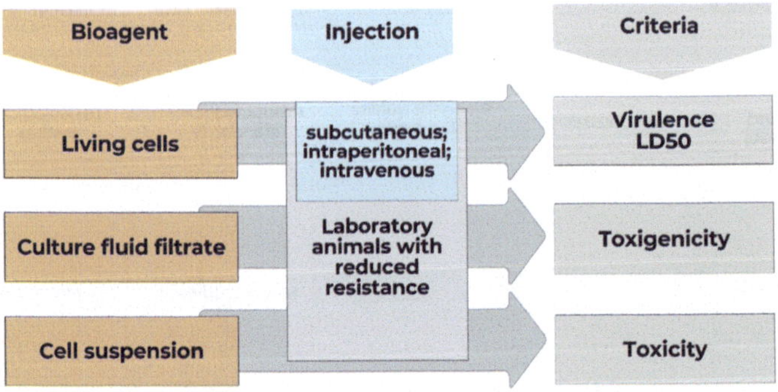

Fig. 2.8 Scheme of hygienic characteristics of the properties of industrial strains

microbial dissemination in internal organs). As a result of the conducted research, the value of TLV is substantiated.

Criteria for the high hazard of the strain (1st hazard class) are:

- LD_{50} value for the insertion of the microorganism into the stomach of mice is less than 10^7, and for intraperitoneal administration, it is less than 10^5 cells per animal.
- Pronounced toxicity—the LD_{50} value for the insertion of a heat-killed culture is less than 10^6 cells per animal.
- Pronounced toxigenicity—the LD_{50} value for intraperitoneal administration of the filtrate is less than 0.5 ml per animal [82].

If the death of animals is not observed, then for the reliability of the assessment of the degree of pathogenicity of the strains, the virulence is determined in a group of weakened animals using experimental loads.

Non virulent strains are tested for the possibility of carrier. This is determined by incubating the cultures for 30 days from experimental animals' skin, mucous membranes, blood, liver, and kidneys. With positive results, the strain is not recommended for use.

When substantiating the TLV of finished forms of preparations containing live microorganisms or spores, in addition to the described studies, the irritant effect of the drug on the skin is assessed, and the threshold of acute and chronic general toxic action is determined.

Based on the received data, the limiting indicator of harmfulness, which is the basis of hygienic rationing of these strains and products, comes to light.

Large-scale studies of industrial non-pathogenic strains and products that comprise the biomass of their inactivated cells have shown that the basis of the manifestations of their adverse effects on the macroorganism is the possibility of sensitization of the organism.

Sensitization as a state of hypersensitivity develops in some people in response to the reintroduction of allergens and depends on the nature and properties of the allergen, its quantity, pathways of penetration and the reactivity of the macroorganism.

The sensitizing effect of microorganisms of producer strains is detected with live culture.

When evaluating the finished forms of preparations for sensitization of animals, use a suspension of preparations (also considering the effect of filler or additives, or solutions of chemical compounds included in the drug), and use live culture (considering the effect of the producer strain).

The reproduction of allergies with repeated reintroduction of the allergen determines the sensitizing effect.

As an example, mice (10–20 individuals) underwent sensitization either intranasally or by an enteral suspension of microorganism cells at a concentration one order of magnitude lower than the LD50. After 5 days, the animals of the study group are tested by introducing into the hind paw 0.05 ml of a suspension of the test microorganism in the same concentration as in sensitization.

After one day, the difference in edema of the two hind legs of the studied and control groups of animals is determined.

When determining the sensitizing effect of the finished form of the preparations for sensitization of animals using intranasal or enteral administration of the drug in concentrations an order of magnitude lower than LD_{50}, as well as inhalation with concentrations selected for Lim_{ac} (threshold concentration for general acute toxicity).Testing is performed by re-administration of the preparations.

To establish the threshold concentration for general chronic toxicity (LimCh), animals sensitized by the suspension of the microorganism under study are subjected for 4 months in special chambers to the inhalation action of an aerosol containing cells of the same microorganism. Daily exposure when setting the TLV in the air of the work zone—4 h, and for the TLV in the air—around the clock.

The threshold of immunotoxic action is determined based on a study of the possible occurrence of the following effects in animals: sensitization, immunization (a sign of antigenicity of the microorganism) and nonspecific immunomodulation (stimulation and immunodeficiency) of the organism. Determination of the dissemination threshold of microorganisms in the internal organs is carried out at the end of the chronic experiment two weeks after its completion (recovery period). For this, blood sampling is performed, and the method of imprinting from the lungs, heart, liver, spleen, and kidneys onto agar-selective media is used. The threshold dose is the minimum dose or concentration. Two weeks after the end of the chronic experiment, at this dose, the studied microorganism is isolated from internal organs, or pathological morphological changes are detected.

The value of the TLV is set based on the limiting threshold of chronic action of the microorganism (immunotoxicity, dysbiosis, and dissemination in internal organs). The stock ratio when determining the TLV in the air of the working space—is 10, and when determining the TLV in atmospheric air—is 100 [83].

The hazard class of microorganisms is determined according to Table 2.1, and the label "A" (allergen) is assigned if the limiting characteristic for regulation is its sensitizing effect.

Producing microorganisms and finished products based on them, classified as hazard class 1, is prohibited in production [84, 85].

1st class—extremely dangerous microorganisms have a pronounced general toxic and allergenic effect.

2nd class—highly dangerous microorganisms can have a strong allergenic and general toxic effect.

3rd class—moderate microorganisms have a weak general toxic and allergenic effect.

4th class—low-risk microorganisms, has virtually no allergenic and general toxic effects.

All industrial strains of microorganisms approved for use are non-pathogenic or opportunistic (except for those used to produce vaccines). According to the normative and technical documentation, they are referred to as the 3rd hazard class (moderate

Table 2.1 Classification of strains of microorganisms by degree of danger

Name of the indicator	Unit of measurement	Standard for hazard class			
		1st	2nd	3rd	4th
Average virulent dose during administration into the stomach	CFU/animal	To 10^7	10^7–10^9	10^9–10^{11}	More than 10^{11}
Average virulent dose during administration into the peritoneum		To 10^5	10^5–10^7	10^7–10^9	More than 10^9
Threshold of allergenic inhalation action	CFU/m^3	Up to 10^2	10^2–10^3	10^3–10^4	More than 10^4
Threshold of chronic inhalation action	CFU/m^3	Up to 10^2	10^3–10^4	10^4–10^5	More than 10^5
Chronic inhalation threshold	CFU/m^3	To 3×10^3	3×10^3–3×10^4	3×10^4–3×10^5	More than 3×10^5
Threshold limit value of microbial strains in workplace air	CFU/m^3	Up to 200	200–2000	2000–20,000	More than 20,000 to 2000
Threshold limit value of microbial strains in atmospheric air		To 20	20–200	200–1000	

individual risk and limited risk for the population as a whole). Such microorganisms, in violation of the technological process and sanitary and hygienic working conditions, can adversely affect the body of workers and lead to microcarriers and allergies.

In addition to the hazard class and TLV of of various pollutants, other indicators are also established to assess their effects on the human organism:

- The value of exceeding the acceptable daily intake (ADI) of pollutant in the body above the threshold limit value (mg/kg body weight).
- The approximate permissible level of the influencing factor in water bodies (mg/dm^3).
- The approximate safe level of the influencing factor in the workplace air and atmospheric air (mg/m^3).
- The approximate permissible level of the influencing factor in the soil (mg/kg).

Since individual pathogenic strains are found among non-pathogenic yeasts of the same species, it is theoretically possible to spontaneously displace an industrial non-pathogenic strain with another virulent strain. Of great importance for ensuring the safe working conditions of staff is the regular microbiological control of the technological process in order to ensure the dominance of production strains of biological agent in these processes.

References

1. Prosser J, Bohannan B, Curtis T et al (2007) The role of ecological theory in microbial ecology. Nat Rev Microbiol 5:384–392. https://doi.org/10.1038/nrmicro1643
2. Mauerhofer LM, Pappenreiter P, Paulik C, Seifert AH, Bernacchi S, Rittmann SKR (2019) Methods for quantification of growth and productivity in anaerobic microbiology and biotechnology. Folia Microbiol 64(3):321–360. https://doi.org/10.1007/s12223-018-0658-4
3. Moreno-Sanchez R, Saavedra E, Rodríguez-Enríquez S, Olín-Sandoval V (2008) Metabolic control analysis: a tool for designing strategies to manipulate metabolic pathways. J Biomed Biotechnol 2008:597913. https://doi.org/10.1155/2008/597913
4. Exponential Phase (2017) ScienceDirect topics. https://www.sciencedirect.com/topics/engineering/exponential-phase
5. Lipson DA (2015) The complex relationship between microbial growth rate and yield and its implications for ecosystem processes. Front Microbiol 6:615. https://doi.org/10.3389/fmicb.2015.00615
6. Palma V, Gutiérrez MS, Vargas O, Parthasarathy R, Navarrete P (2022) Methods to evaluate bacterial motility and its role in bacterial-host interactions. Microorganisms 10(3):563. https://doi.org/10.3390/microorganisms10030563
7. Vadillo Rodríguez V (2004) Macroscopic and microscopic approaches toward bacterial adhesion. University of Groningen. https://pure.rug.nl/ws/portalfiles/portal/14633966/c1.pdf
8. Ramirez J, Meraz M, Vernon-Carter EJ (2019) A theoretical derivation of the Monod equation with a kinetics sense. Biochem Eng J 150:107305. https://doi.org/10.1016/j.bej.2019.107305
9. Strigul N, Dette H, Melas VB (2009) A practical guide for optimal designs of experiments in the Monod model. Environ Model Softw 24(9):1019–1026. https://doi.org/10.1016/j.envsoft.2009.02.006
10. Longhi DA, Dalcanton F, de Aragão GMF, Carciofi BAM, Laurindo JB (2017) Microbial growth models: a general mathematical approach to obtain μmax and λ parameters from sigmoidal empirical primary models. Braz J Chem Eng 34(2):369–375. https://doi.org/10.1590/0104-6632.20170342s20150533
11. Kumar J, Ramlal A, Mallick D, Mishra V (2021) An overview of some biopesticides and their importance in plant protection for commercial acceptance. Plants 10(6):1185. https://doi.org/10.3390/plants10061185
12. Recent Advancement in Microbial Biotechnology (2021) Biopesticide. ScienceDirect. https://www.sciencedirect.com/topics/agricultural-and-biological-sciences/biopesticide
13. Sumbul A, Ansari RA, Rizvi R, Mahmood I (2020) Azotobacter: a potential bio-fertilizer for soil and plant health management. Saudi J Biol Sci 27(12):3634–3640. https://doi.org/10.1016/j.sjbs.2020.08.004
14. George E, Kothari SK, Li XL, Weber E, Marschner H (1994) VA mycorrhiza: benefits to crop plant growth and costs. In: Muehlbauer FJ, Kaiser WJ (eds) Expanding the production and use of cool season food legumes. Current plant science and biotechnology in agriculture, vol 19. Springer, Dordrecht. https://doi.org/10.1007/978-94-011-0798-3_51
15. Kalayu G (2019) Phosphate solubilizing microorganisms: promising approach as biofertilizers. Int J Agron 2019:4917256. https://doi.org/10.1155/2019/4917256
16. Massot F, Bernard N, Alvarez LMM et al (2022) Microbial associations for bioremediation. What does "microbial consortia" mean? Appl Microbiol Biotechnol 106:2283–2297. https://doi.org/10.1007/s00253-022-11864-8
17. Lapage SP, Sneath PHA, Lessel EF, Skerman VBD, Seeliger HPR, Clark WA (1992) International code of nomenclature of bacteria: bacteriological code, 1990 revision. ASM Press, Washington, DC. ISBN: 1-55581-039-X
18. Arahah DR, Bull CT, Busse H-J et al (2023) Guidelines for interpreting the International Code of Nomenclature of Prokaryotes and for preparing a request for an opinion. Int J Syst Evol Microbiol 73(1):005782. https://doi.org/10.1099/ijsem.0.005782
19. Bacaër N (2011) Verhulst and the logistic equation (1838). In: A short history of mathematical population dynamics. Springer, London. https://doi.org/10.1007/978-0-85729-115-8_6

20. Noell SE, Brennan E, Washburn Q, Davis EW 2nd, Hellweger FL, Giovannoni SJ (2023) Differences in the regulatory strategies of marine oligotrophs and copiotrophs reflect differences in motility. Environ Microbiol 25(7):1265–1280. https://doi.org/10.1111/1462-2920.16357
21. Ho A, Di Lonardo DP, Bodelier PLE (2017) Revisiting life strategy concepts in environmental microbial ecology. FEMS Microbiol Ecol 93(3):fix006. https://doi.org/10.1093/femsec/fix006
22. Wilkins MR (2022) Chapter 14—Factors influencing microbial growth and its kinetics. In: Current developments in biotechnology and bioengineering: advances in bioprocess engineering. Elsevier, pp 373–395. https://doi.org/10.1016/B978-0-323-91167-2.00014-9
23. Sies H, Mailloux RJ, Jakob U (2024) Fundamentals of redox regulation in biology. Nat Rev Mol Cell Biol 25:701–719. https://doi.org/10.1038/s41580-024-00730-2
24. Liebensteiner MG, Tsesmetzis N, Stams AJM, Lomans BP (2014) Microbial redox processes in deep subsurface environments and the potential application of (per)chlorate in oil reservoirs. Front Microbiol 5:428. https://doi.org/10.3389/fmicb.2014.00428
25. Liu L-Y, Wang X, Dang C-C, Zhao Z-C, Xing D-F, Liu B-F, Ren N-Q, Xie G-J (2024) Anaerobic ammonium oxidation coupled with sulfate reduction links nitrogen with sulfur cycle. Biores Technol 403:130903. https://doi.org/10.1016/j.biortech.2024.130903
26. Liu L-Y, Xie G-J, Xing D-F, Liu B-F, Ding J, Cao G-L, Ren N-Q (2021) Sulfate dependent ammonium oxidation: a microbial process linked nitrogen with sulfur cycle and potential application. Environ Res 192:110282. https://doi.org/10.1016/j.envres.2020.110282
27. Popovic M (2019) Thermodynamic properties of microorganisms: determination and analysis of enthalpy, entropy, and Gibbs free energy of biomass, cells and colonies of 32 microorganism species. Heliyon. 5(6):e01950. https://doi.org/10.1016/j.heliyon.2019.e01950
28. Popovic M, Šekularac G, Stevanović M (2024) Thermodynamics of microbial consortia: enthalpies and Gibbs energies of microorganism live matter and macromolecules of *E. coli*, *G. oxydans*, *P. fluorescens*, *S. thermophilus* and *P. chrysogenum*. J Biotechnol 379:6–17. https://doi.org/10.1016/j.jbiotec.2023.11.001
29. Advances in Applied Microbiology (2021) Syntrophy. Agric Biol Sci. https://www.sciencedirect.com/topics/agricultural-and-biological-sciences/syntrophy
30. Weng N, Singh A, Ohlsson JA, Dolfing J, Westerholm M (2024) Catabolism and interactions of syntrophic propionate- and acetate oxidizing microorganisms under mesophilic, high-ammonia conditions. Front Microbiol 15:1389257. https://doi.org/10.3389/fmicb.2024.1389257
31. Crognale S, Casentini B, Amalfitano S, Fazi S, Petruccioli M, Rossetti S (2019) Biological As(III) oxidation in biofilters by using native groundwater microorganisms. Sci Total Environ 651(Pt 1):93–102. https://doi.org/10.1016/j.scitotenv.2018.09.176
32. Nazina TN, Sokolova DS, Shestakova NM et al (2005) The phylogenetic diversity of aerobic organotrophic bacteria from the Dagang high-temperature oil field. Microbiology 74:343–351. https://doi.org/10.1007/s11021-005-0073-0
33. Protein Byproducts (2016) Biochemical oxygen demand. Agric Biol Sci. https://www.sciencedirect.com/topics/agricultural-and-biological-sciences/biochemical-oxygen-demand
34. Material Selection for Thermoplastic Parts (2016) Biochemical oxygen demand. Engineering. https://www.sciencedirect.com/topics/engineering/biochemical-oxygen-demand
35. Costa Almeida OA, Oliveira de Araujo N, Mulato ATN, Persinoti GF, Sforça ML, Calderan-Rodrigues MJ, Oliveira JVC (2023) Bacterial volatile organic compounds (VOCs) promote growth and induce metabolic changes in rice. Front Plant Sci 13:1056082. https://doi.org/10.3389/fpls.2022.1056082
36. Chandrasekaran M, Paramasivan M, Sahayarayan JJ (2023) Microbial volatile organic compounds: an alternative for chemical fertilizers in sustainable agriculture development. Microorganisms 11:42. https://doi.org/10.3390/microorganisms11010042
37. Encyclopedia of Fish Physiology (2011) Hydrolysis. Immunol Microbiol. https://www.sciencedirect.com/topics/immunology-and-microbiology/hydrolysis
38. Amin FR, Khalid H, El-Mashad HM, Chen C, Liu G, Zhang R (2021) Functions of bacteria and archaea participating in the bioconversion of organic waste for methane production. Sci Total Environ 763:143007. https://doi.org/10.1016/j.scitotenv.2020.143007

39. Harindintwali JD, Xiang L, Wang F, Chang SX, Zhao Z, Mei Z, Tiedje JM (2022) Syntrophy of bacteria and archaea in the anaerobic catabolism of hydrocarbon contaminants. Crit Rev Environ Sci Technol 53(13):1331–1357. https://doi.org/10.1080/10643389.2022.2134702
40. Das T, Panchanadikar V, Chaudhury GR (1997) Bio-oxidation of iron using *Thiobacillus ferrooxidans*. World J Microbiol Biotechnol 14:297–298. https://doi.org/10.1023/A:1008867104672
41. Bai X, Dinkla IJT, Muyzer G (2022) Microbial ecology of biofiltration used for producing safe drinking water. Appl Microbiol Biotechnol 106(13–16):4813–4829. https://doi.org/10.1007/s00253-022-12013-x
42. Saravanan A, Karishma S, Senthil KP, Rangasamy G (2023) Biodegradation of oil-contaminated aqueous ecosystem using an immobilized fungi biomass and kinetic study. Environ Res 220:115252. https://doi.org/10.1016/j.envres.2023.115252
43. Zahid S, Aziz SH, Jazza SH, Dageem HN, Banoon SR, Balboul BA, Abdelzaher MA (2024) Bacterial biodegradation of oil-contaminated soil for pollutant abatement contributing to achieving sustainable development goals: a comprehensive review. Results Eng. 22:102083. https://doi.org/10.1016/j.rineng.2024.102083
44. Water Reclamation and Sustainability (2014) Biological wastewater treatment. Chem Eng. https://www.sciencedirect.com/topics/chemical-engineering/biological-wastewater-treatment
45. Sravan JS, Matsakas L, Sarkar O (2024) Advances in biological wastewater treatment processes: focus on low-carbon energy and resource recovery in biorefinery context. Bioengineering 11(3):281. https://doi.org/10.3390/bioengineering11030281
46. Urrea JL, Collado S, Laca A, Díaz M (2014) Wet oxidation of activated sludge: transformations and mechanisms. J Environ Manage 146:251–259. https://doi.org/10.1016/j.jenvman.2014.07.043
47. Narayanan CM, Narayan V (2019) Biological wastewater treatment and bioreactor design: a review. Sustain Environ Res 29:33. https://doi.org/10.1186/s42834-019-0036-1
48. Encyclopedia of Neuroscience (2009) Michaelis-Menten kinetics. Immunol Microbiol. https://www.sciencedirect.com/topics/immunology-and-microbiology/michaelis-menten-kinetics
49. Biochimica et Biophysica Acta (BBA)—Proteins and Proteomics (2011) Denitrification. Biochem Gen Mol Biol. https://www.sciencedirect.com/topics/biochemistry-genetics-and-molecular-biology/denitrification
50. Bioresource Technology (2020) Sulphate reducer. Earth Planet Sci. https://www.sciencedirect.com/topics/earth-and-planetary-sciences/sulphate-reducer
51. Nagakura T, Schubert F, Wagner D, Kallmeyer J, IODP Exp. 385 Shipboard Scientific Party (2022) Biological sulfate reduction in deep subseafloor sediment of Guaymas Basin. Front Microbiol 13:845250. https://doi.org/10.3389/fmicb.2022.845250
52. Yang Z, Muhayodin F, Larsen OC, Miao H, Xue B, Rotter VS (2021) A review of composting process models of organic solid waste with a focus on the fates of C, N, P, and K. Processes 9:473. https://doi.org/10.3390/pr9030473
53. Sidelko R, Janowska B, Walendzik B, Siebielska I (2010) Two composting phases running in different process conditions timing relationship. Biores Technol 101(17):6692–6698. https://doi.org/10.1016/j.biortech.2010.03.092
54. Yin J, Xie M, Yu X, Feng H, Wang M, Zhang Y, Chen T (2024) A review of the definition, influencing factors, and mechanisms of rapid composting of organic waste. Environ Pollut 342:123125. https://doi.org/10.1016/j.envpol.2023.123125
55. Johnson DB (2014) Biomining—biotechnologies for extracting and recovering metals from ores and waste materials. Curr Opin Biotechnol 30:24–31. https://doi.org/10.1016/j.copbio.2014.04.008
56. Rendón-Castrillón L, Ramírez-Carmona M, Ocampo-López C, Gómez-Arroyave L (2023) Bioleaching techniques for sustainable recovery of metals from solid matrices. Sustainability 15:10222. https://doi.org/10.3390/su151310222
57. Tezyapar KI, Kremser K, Wagland ST et al (2023) Bioleaching metal-bearing wastes and by-products for resource recovery: a review. Environ Chem Lett 21:3329–3350. https://doi.org/10.1007/s10311-023-01611-4

58. Rohrmann GF (2019) Baculovirus molecular biology, 4th edn. National Center for Biotechnology Information (US), Bethesda, MD
59. Katsimpouras C, Stephanopoulos G (2021) Enzymes in biotechnology: critical platform technologies for bioprocess development. Curr Opin Biotechnol 69:91–102. https://doi.org/10.1016/j.copbio.2020.12.003
60. Khan MR (2021) Immobilized enzymes: a comprehensive review. Bull Natl Res Centre 45:207. https://doi.org/10.1186/s42269-021-00649-0
61. Basso A, Serban S (2019) Industrial applications of immobilized enzymes—a review. Mol Catal 479:110607. https://doi.org/10.1016/j.mcat.2019.110607
62. Maghraby YR, El-Shabasy RM, Ibrahim AH, Azzazy HME (2023) Enzyme immobilization technologies and industrial applications. ACS Omega 8(6):5184–5196. https://doi.org/10.1021/acsomega.2c07560
63. Narayanan M, Ali SS, El-Sheekh M (2023) A comprehensive review on the potential of microbial enzymes in multipollutant bioremediation: mechanisms, challenges, and future prospects. J Environ Manage 334:117532. https://doi.org/10.1016/j.jenvman.2023.117532
64. Value-Addition in Food Products and Processing Through Enzyme Technology (2022) Enzyme substrate complex. Biochem Gen Biol. https://www.sciencedirect.com/topics/biochemistry-genetics-and-molecular-biology/enzyme-substrate-complex
65. McCullagh Lab (2024) Enzyme kinetics: the Michaelis-Menten mechanism. Chem Lessons Jupyter Notebooks. https://mccullaghlab.github.io/physical_chemistry/kinetics/Enzyme_Kinetics_Michaelis_Menten.html
66. Yan A, Wang Y, Tan SN, Yusof MLM, Ghosh S, Chen Z (2020) Phytoremediation: a promising approach for revegetation of heavy metal-polluted land. Front Plant Sci 11:359. https://doi.org/10.3389/fpls.2020.00359
67. Kafle A, Timilsina A, Gautam A, Adhikari K, Bhattarai A, Aryal N (2022) Phytoremediation: mechanisms, plant selection and enhancement by natural and synthetic agents. Environ Adv 8:100203. https://doi.org/10.1016/j.envadv.2022.100203
68. Pivetz BE (2001) Phytoremediation of contaminated soil and ground water at hazardous waste sites. EPA Ground Water Issue (EPA/540/S-01/500). Office of Research and Development, U.S. Environmental Protection Agency
69. Guo Y, Wen L, Zhao X, Xing C, Huang R (2024) Industrial hemp (*Cannabis sativa* L.) can utilize and remediate soil strongly contaminated with Cu, As, Cd, and Pb by phytoattenuation. Chemosphere 358:142199. https://doi.org/10.1016/j.chemosphere.2024.142199
70. Shah V, Daverey A (2020) Phytoremediation: a multidisciplinary approach to clean up heavy metal contaminated soil. Environ Technol Innov 18:100774. https://doi.org/10.1016/j.eti.2020.100774
71. Shehata SM, Badawy RK, Aboulsoud YIE (2019) Phytoremediation of some heavy metals in contaminated soil. Bull Natl Res Centre 43:189. https://doi.org/10.1186/s42269-019-0193-3
72. Khan AU, Khan AN, Waris A, Ilyas M, Zamel D (2022) Phytoremediation of pollutants from wastewater: a concise review. Open Life Sci 17(1):488–496. https://doi.org/10.1515/biol-2022-0056
73. Kristanti RA, Hadibarata T (2023) Phytoremediation of contaminated water using aquatic plants, its mechanism and enhancement. Curr Opin Environ Sci Health 32:100451. https://doi.org/10.1016/j.coesh.2023.100451
74. Park JK, Oh K (2023) Advancements in phytoremediation research for soil and water resources: harnessing plant power for environmental cleanup. Sustainability 15:13901. https://doi.org/10.3390/su151813901
75. Lapan O, Mikhieyev O, Madzhid S, Cherniak L, Maksimenko O (2022) Development of the hydrophytic structure of the bioplateau type for the purification of water bodies from 137Cs. Ecol Balkanica 14(1):1–9
76. Lapan OV, Kudryavtseva DO (2018) Experience in the use of hydro-fit structures in Ukraine. In: Modern hydroecology: the role of scientific research in solving current problems: proceedings of the 5th scientific and practical conference for young scientists. Kyiv, pp 31–32

77. Morais Junior WG, Gorgich M, Corrêa PS, Martins AA, Mata TM, Caetano NS (2020) Microalgae for biotechnological applications: cultivation, harvesting and biomass processing. Aquaculture 528:735562. https://doi.org/10.1016/j.aquaculture.2020.735562
78. Alvarez AL, Weyers SL, Goemann HM, Peyton BM, Gardner RD (2021) Microalgae, soil and plants: a critical review of microalgae as renewable resources for agriculture. Algal Res 54:102200. https://doi.org/10.1016/j.algal.2021.102200
79. Gonçalves J, Freitas J, Fernandes I, Silva P (2023) Microalgae as biofertilizers: a sustainable way to improve soil fertility and plant growth. Sustainability 15:12413. https://doi.org/10.3390/su151612413
80. Ramakrishnan B, Maddela NR, Venkateswarlu K, Meghara M (2023) Potential of microalgae and cyanobacteria to improve soil health and agricultural productivity: a critical view. Environ Sci Adv 2:586–611. https://doi.org/10.1039/D2VA00158F
81. Egli T (2015) Microbial growth and physiology: a call for better craftsmanship. Front Microbiol 6:287. https://doi.org/10.3389/fmicb.2015.00287
82. Scientific Committees (2024) LD50: lethal dose 50, median lethal dose. University of Guelph. https://ec.europa.eu/health/scientific_committees/opinions_layman/en/tooth-whiteners/glossary/jkl/ld50.htm. Accessed 4 June 2024
83. Van Rossum T, Ferretti P, Maistrenko OM, Bork P (2020) Diversity within species: interpreting strains in microbiomes. Nat Rev Microbiol 18(9):491–506. https://doi.org/10.1038/s41579-020-0368-1
84. de la Cal MA, Cerdà E, Abella A, Garcia-Hierro P (2011) Classification of microorganisms according to their pathogenicity. In: van Saene H, Silvestri L, de la Cal M, Gullo A (eds) Infection control in the intensive care unit. Springer, Milano. https://doi.org/10.1007/978-88-470-1601-9_3
85. Shanson DC (1989) Classification and pathogenicity of microbes. In: Microbiology in clinical practice, pp 3–32. https://doi.org/10.1016/B978-0-7236-1403-6.50010-7

Chapter 3
Substrates of Environmental Biotechnologies

Abstract In general, the substrates of biotechnology can be very different depending on the sphere of production and application. However, a particular area of biotechnology implementation in environmental protection is the introduction of waste recycling. Therefore, this chapter deals with the possibilities of waste recycling in biotechnology, focusing on reducing the negative impact on the environment. Thus, the approach aims to change the attitude towards waste, not as unnecessary material accumulating in the environment and harming it. It also considers waste as a sustainable raw material and biomass, which is necessary for biological agents as a substrate for their growth.

3.1 Waste as a Substrate

Substrates and media used in environmental biotechnology are very different; their range constantly expands (Table 3.1) [1].

With the development of industrial processes, there is an accumulation of new types of waste that can be neutralized and converted into useful products by biotechnology methods [2]. On the one hand, biotechnological industries, which are developing rapidly, face the problem of depletion of traditional raw materials, so there is a need to expand the raw material base; on the other—increasing the amount of accumulated waste, necessitates the development of non-traditional, including biotechnological methods.

Thus, *in environmental biotechnology, the substrate for the development of biological agent is primarily the waste of production and consumption, which is a secondary raw material base and object of processing.* This trend is based on implementing the principles of low waste in terms of greening technological processes [3, 4].

Y. Chernysh et al., *Sustainable Biotechnology and Environmental Protection*, Green Energy and Technology, https://doi.org/10.1007/978-3-031-89436-7_3

Table 3.1 Groups of substrates, biological agents and products formed in biotechnological processes

Substrate	Biological agent	Product
Hydrol; hydrolysates of wood waste; stillage; oil paraffins; surfactants; intermediates: precursors of biotransformations	Microorganisms, plant and animal cells; extracellular products: enzymes, coenzymes Immobilized cells of microorganisms and plants	Biofertilizers and bioinsecticides, microbial biomass, vaccines; pharmaceuticals; hormones; polysaccharides; unicellular protein
Whey; green plant biomass; agriculture and forestry industry waste; household waste, wastewater; municipal sewage sludge; fruit and vegetable processing waste; metallurgical and chemical industry waste; sulfite liquor	Grouping of microorganisms	Organic acids; food products; antibiotics; amino acids; alcohols, organic solvents; enzymes, vitamins; biogas, organic fertilizer; metals, non-metals; elemental sulfur; feed yeast

3.2 Principles of Preparation of Nutrient Media for Inoculant Strains

All living cells need exogenous power sources contained in nutrient media. The nutrient medium provides the viability, growth and development of a biological agent, which can then be used as a starter culture (inoculant) to intensify the process of treatment and disposal of waste.

Inoculation—the introduction of living organisms into different environments. A constant component of nutrient media is water, which all living cells need. The supply of cells with oxygen and hydrogen is carried out by water. Nutrients form true (mineral salts, sugars, amino acids, carboxylic acids, alcohols, aldehydes) or colloidal (proteins, lipids, inorganic compounds) solutions. Some components of the nutrient media, which are in a solid state, can either form a bottom sediment, or be evenly distributed throughout the volume in the form of a suspension, or float on the solution's surface (coal particles)—liquid hydrocarbons when introduced into water form an immiscible fraction.

The nutrient medium must have all the elements necessary to build living cell components in an accessible form. *Mineral elements* that are part of nutrient media and necessary for biological agent growth are divided into macro- and micronutrients. Their concentration in the culture medium depends on the biological agent biology and the biotechnological process's tasks. Among *macronutrients*, nitrogen holds the primary position, as the requirements of biological objects for it are an order of magnitude higher than for other elements (phosphorus, sulfur, potassium and magnesium). Thus, the concentration of macronutrients in the medium (K, Mg, P, S) is usually about 10^{-3} to 10^{-4} M. Salts (sulfates or ammonium phosphates) are used as their sources. Potassium concentration changes little during growth. Usually, the K introduced with phosphates is sufficient. In experimental media, phosphate concentrations are primarily regulated to maintain pH rather than to meet the biological agent's needs. 100 mg/l of potassium phosphate is enough for the culture's stationary growth phase, even at high density.

Nitrogenous substrates for the preparation of nutrient media are mainly proteins of animal and vegetable origin. It is important to note that the sources of nitrogen organisms differ sharply. There are three main sources of bound nitrogen: ammonium salts, nitrates, and organic nitrogen, which can be either in the composition of yeast extract or in the form of glutamate, glutamine, aspartate, asparagine, and urea. For anaerobes, it is better to use ammonium salt; for aerobes, it is nitrates.

Not only the appearance but also the concentration of nitrogen in the culture media is essential in providing nitrogen to the biological agent because the change in the C:N ratio affects the growth rate of the agent's metabolism and causes the over synthesis of a number of target products (amino acids, polysaccharides, etc.).

The need for micronutrients is small, but their concentration in the media is significantly lower—10^{-6} to 10^{-8} M. Therefore, micronutrients are often deliberately not introduced into the media because their impurities in basic salts and water meet the needs of biological agents. The problem with trace elements is to keep them in solution at a given pH. This is usually achieved by using complexing agents—nitrile trioacetic acid or EDTA (complexone). Iron citrate is used for iron, which requires a relatively large amount. The above trace elements are associated with certain functions in the cell. Cobalt is involved in the reactions of vitamin B12, molybdenum—in nitrogen metabolism, copper—in a number of oxidative enzymes, nickel—in hydrogen genesis, etc. The absence or lack of these trace elements in the media leads to the selective growth of other organisms.

In addition to the main energy and plastic components, nutrient media may contain so-called *growth factors.* These are organic compounds (vitamins, amino acids, purine and pyrimidine bases, etc.) that auxotrophic cells need but cannot synthesize. The absence of such substances leads to disruption of metabolic processes and cessation of cell growth. In addition to pure individual substances of such nature as growth supplements, in practice often used corn or yeast extract, potato juice, malt extract, grain waste, and dairy industry by-products.

The stimulating effect of these growth factors largely depends on the individual properties of the applied biological agent, the composition of the basic medium, fermentation conditions, and others. Adding growth factors can increase the yield of the target product, such as increasing the amount of enzymes tenfold [5]. The necessary components of nutrient media can be gases: soluble (NH_3, H_2S), moderately (CO_2) or poorly (N_2, O_2, H_2, CH_4) soluble in water.

Carbon is an integral part of all organic compounds, and *its sources are* numerous and diverse: most often sugars, polyhydric alcohols and organic acids.

Nutrient media may have an indeterminate composition, i.e. contain nutrients (vegetable, animal or microbial origin), such as meat extract, yeast extract, corn flour, seaweed and others. Such nutrient media are called *natural.* At present, there is a growing interest of biotechnologists in natural renewable resources - products of photosynthesis and bioresources of the oceans.

As *sources of carbon* and energy in biotechnological processes the main used are natural complex media of indeterminate content (wastes of various industries, processed products, wastewater components and other organic wastes), which, in addition to carbon compounds, also contain mineral elements and growth factors. Cellulose, hydrolysates of polysaccharides and wood, are quite widely included in the category of biotechnological substrates.

The media is also prepared from purely chemical compounds in predetermined proportions. These are the so-called *synthetic media*. *Semi-synthetic nutrient media* that combine components of both natural and synthetic media are also commonly used. Culture media of these types have both advantages and disadvantages. From an economic point of view, the most appropriate is the use of natural, cheaper raw materials than substances in pure form obtained by synthetic means. However, only well-defined media parameters allow the accurate monitoring and regulation of the passage of culture media processes, seeking to optimize them. A compromise approach is using semi-synthetic media, which, along with compounds of known chemical nature, include nutrients.

Traditionally, the composition of the nutrient medium optimal for the biotechnological process is determined by the method of long-term empirical selection, during which the qualitative and quantitative composition of the medium is determined in the first stages. Many attempts have been made to standardize the composition of media from the standpoint of physiology and biochemistry of the target organism. However, since nutrient needs are species- and even strain-specific, in each case, it is necessary to optimize the composition of the media for a particular target organism. Over the last 20–25 years, the mathematical method of experiment planning and mathematical modelling of biotechnological processes has been increasingly used. This allows a reasonable approach to the design of nutrient media to make them economical [3].

References

1. Astudillo Á, Rubilar O, Briceño G, Diez MC, Schalchli H (2023) Advances in agroindustrial waste as a substrate for obtaining eco-friendly microbial products. Sustainability 15:3467. https://doi.org/10.3390/su15043467
2. Danylov DV, Chubur VS, Chernysh YY, Yakhnenko OM (2020) Bioenergy waste recycling: modelling of development trends. Man Environ Issues Neoecol **2020**(34):141–152
3. Stikane A, Baumanis MR, Muiznieks R, Stalidzans E (2023) Impact of waste as a substrate on biomass formation, and optimization of spent microbial biomass re-use by sustainable metabolic engineering. Fermentation 9:531. https://doi.org/10.3390/fermentation9060531
4. Volkmar M, Maus AL, Weisbrodt M, Bohlender J, Langsdorf A, Holtmann D, Ulber R (2023) Municipal green waste as substrate for the microbial production of platform chemicals. Bioresour Bioprocess 10(1):43. https://doi.org/10.1186/s40643-023-00663-2
5. World Market for Enzyme Production (2016) Access mode. http://www.nauka.kz/biol_med/razd4/mir_rinok_proizvods_enzimov.php

Chapter 4
Aerobic and Anaerobic Processes Waste Fermentation

Abstract The processes of wastewater and sludge treatment and waste detoxification in biotechnology can be divided into aerobic and anaerobic. Each subtype of biotechnology has its peculiarities in achieving environmental safety indicators. In this chapter, the biochemical basis of fermentation processes in oxygen and oxygen-free modes, as well as the necessary technological solutions and technical features, are discussed.

4.1 Biochemical Processes of Oxidation of Organic Substances in Aerobic and Anaerobic Conditions

The basis of biological treatment processes is the ability of bacteria to use various organic and mineral substances of wastewater as food in the process of their life cycle. Bacteria have a wide variety of physiological capabilities in terms of nutrients and environmental conditions, which allow the removal of almost any organic compounds from wastewater.

Many biochemical reactions take place simultaneously in bacterial cells. Enzymes that accelerate biochemical reactions have high catalytic activity, i.e. effectively reducing the activation energy required for the reaction as they promote the formation of intermediate products that require less energy.

It is believed that the mechanism of action of enzymes is as follows: the enzyme protein binds at one or more points with the substrate molecule, forming an enzyme–substrate complex that weakens the intramolecular bonds of the substrate. The Michaelis–Menten equation theoretically substantiated the formation of the complex; later, this reaction was confirmed experimentally. As a result, the well-known Michaelis–Menten equation was obtained:

$$\rho = \frac{\rho_{max} L}{K_c + L}, \tag{4.1}$$

Y. Chernysh et al., *Sustainable Biotechnology and Environmental Protection*, Green Energy and Technology, https://doi.org/10.1007/978-3-031-89436-7_4

where ρ is the specific rate of the enzymatic reaction; ρ_{max}—maximum reaction rate in the absence of substrate limitation; K_c—saturation constant; L is the concentration of the substrate.

The rate of the enzymatic reaction increases by increasing substrate concentration to a certain limit.

Enzymes are produced by the cell according to its needs, and their content can vary significantly. Enzymes have a high specificity, and their activity depends on various factors (temperature, pH, nutrient composition, and the presence of toxic substances).

Bacterial decomposition of organic matter can occur under anaerobic and aerobic conditions.

Aerobic conversion reactions of organic matter

Organic substances under aerobic conditions can

- to be transformed into corresponding monomers (simple organic substances).
- oxidize to carbon dioxide and various mineral elements.
- do not undergo transformation (inert and sparingly soluble organic substances).

Organic matter in wastewater has a general appearance $C_{18}H_{19}O_9N$. If microorganisms oxidize this substance to carbon dioxide, the total reaction can be represented by the following chemical equations:

$$C_{18}H_{19}O_9N + 17,5O_2 + H^+ \rightarrow 18CO_2 + 8H_2O + NH_4^+ \text{(without nitrification)}, \quad (4.2)$$

$$C_{18}H_{19}O_9N + 19,5O_2 \rightarrow 18CO_2 + 9H_2O + NO_3^- \text{(with nitrification)}. \quad (4.3)$$

The total energy yield of the oxidation reaction of organic matter by aerobic heterogeneous microorganisms is written in the form:

$$\frac{1}{70}C_{18}H_{19}O_9N + \frac{1}{4}O_2 \rightarrow \frac{17}{70}CO_2 + \frac{1}{70}HCO_3^- + \frac{1}{70}NH_4^+ + \frac{1}{10}H_2O,$$
$$\Delta G^\circ(W) = -110\,\text{kJ/el.eq.} \quad (4.4)$$

In obligate aerobes, biological oxidation reactions are carried out with the participation of oxygen and the release of large amounts of energy. Thus, glucose oxidation under aerobic conditions is accompanied by the release of 688.5 kcal.

Consider in more detail the processes of nitrification and denitrification, which are actively used in environmental biotechnologies for wastewater treatment.

In domestic wastewater, nitrogen is the main part of organic matter, which is the end product of nitrogen metabolism in the human body. Ammonia is released in the proteolytic pathway during the decomposition of nitrogenous compounds, which can inhibit the process of aerobic destruction. 80–90% of all nitrogenous substances are present in the form of ammonia or urea in domestic wastewater.

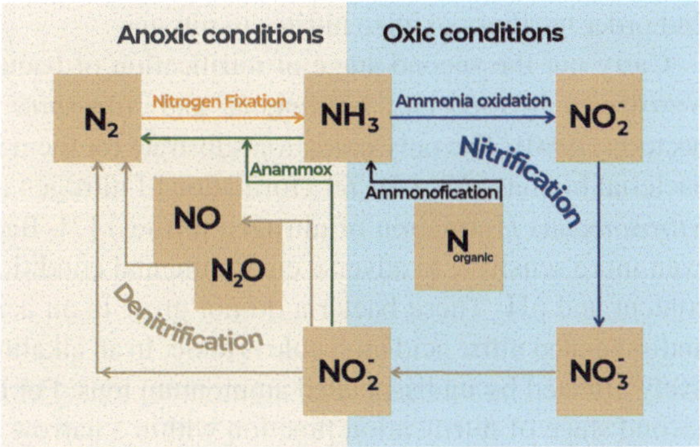

Fig. 4.1 Closed redox cycle of nitrogen: physiological groups of microorganisms

Ammonification is the bacterial conversion of organic nitrogen compounds into inorganic forms, the main of which is ammonia, which accumulates in the process of deamination due to proteolysis of proteins of plant and animal origin, carried out by heterotrophic putrefying (ammonifying) bacteria in the sewer network. In addition to ammonia, phosphorus and hydrogen sulfide are formed due to ammonification [1]. Low temperatures (less than 10 °C) and acidic pH hinder this process.

Figure 4.1 shows a diagram of a closed nitrogen cycle. It should be noted that the *diversity of physiological groups of microorganisms in the carbon and nitrogen cycles indicates that they form integrated flows of elements.* Thus, a complete cycle of conversion of organic matter in the aerobic group is formed.

Nitrification is a complex multistage process. The first stage of nitrification, the oxidation of ammonium salts into nitrites, which carry out nitrifiers of the 1st order, occurs by the equation

$$NH_3 + 5O_2 \rightarrow 2H^+ + 2NO_2 + H_2O. \quad (4.5)$$

The nitrification process is carried out as a result of the vital and functional activity of nitrifying bacteria, which belong to chemosynthetic autotrophs; the presence of organic compounds in the environment has a detrimental effect on development, so the nitrification of ammonium nitrogen begins in aeration tanks only after almost complete oxidation of carbon-containing compounds, characterized by BOD.

While there is an excess of organic matter and intensively developing heterotrophic bacteria—competitors of nitrifiers for ammonia in the processes of constructive metabolism, nitrification is suppressed. In addition, heterotrophic bacteria intensively absorb, as already mentioned, the oxygen needed by nitrifiers. After the organic matter is mineralized, conditions are created for the development of bacteria—the causative agents of the first nitrification phase, carried out by bacteria of several genera. The main ones are *Nitrosomonas, Nitrosocystis, Nitrosospira, Nitrosococcus, Nitrosolobus* and *Nitrosovibrio*. The appearance of nitrites in treated wastewater indicates that the bulk of organic matter is already mineralized (except for processes in irrigation fields, where they occur in parallel), and nitrification has begun.

2nd order nitrifiers oxidize nitrites to nitrates

Carry out the second stage of nitrification of bacteria of the genus *Nitrobacter, Nitrospina, Nitrococcus, Nitrocystis,* and *Nitrospira* (Short determinant of Bergi bacteria, 1980). The only oxidation substrate for them is nitrites. The growth rates of bacterial biomass *Nitrobacter* (formation of nitrogen nitrates) are much higher than *Nitrosomonas* (formation of nitrogen nitrites) [2]. Bacteria in the second stage are even more sensitive to adverse environmental conditions, such as dissolved oxygen content and pH. These bacteria do not grow in an acidic environment because the undissociated nitric acid molecule is toxic. In an alkaline environment, they are negatively affected by undissociated ammonium ions. For this reason, the bacteria of the second stage of nitrification function within a narrow range of neutral pH values of 7.0–7.6.

The denitrification process proceeds as follows:

$$\frac{1}{3}NO_3^- + \frac{6}{3}H^+ + e \rightarrow \frac{1}{10}N_2 + \frac{3}{5}H_2O. \tag{4.6}$$

As we can see from the equation, the electron acceptor (nitrate) is reduced to molecular nitrogen.

Thus, we see a complex structure with self-regulation in the microbial group of aerobes, which can be effectively used in biological wastewater treatment methods and environmental technologies in general.

The main processes used in biological treatment are aerobic, in which organic matter is eventually oxidized to carbon dioxide and water. Cells receive biologically useful energy due to enzymatic reactions, the process by which electrons pass from one energy level to another. For most organisms, the ultimate electron acceptor is oxygen. The transfer of oxygen electrons occurs with the participation of the electron transfer system, which sequentially transfers them to the various components of the system and eventually attaches to oxygen, activating it. Activated oxygen reacts with an ionized hydrogen atom to form water or hydrogen peroxide. In the process of enzymatic reactions, the energy of electrons is bound in the macroergic bonds of ATP.

A reduction reaction accompanies any oxidative reaction. Electron transfer through the electron transfer system, which occurs through a series of successive redox reactions, is biological oxidation.

For the formation and constant accumulation of energy in the process of oxidative metabolism under aerobic conditions, it is extremely important to supply the cell with oxygen and its concentration in the environment.

Increasing the overall rate of oxidation of organic matter in wastewater can be done by increasing the biomass of microorganisms involved in the process. The proportionality between the rate of oxidation and the biomass of microorganisms exists to a certain extent, above which the specific rate of oxidation is significantly reduced due to the deterioration of mass transfer, interspecific and intraspecific competition, increasing the concentration of metabolic products. In this case, the magnitude of the inhibitory factor is proportional to the biomass of microorganisms.

When amending the Michaelis–Menten Eq. (4.1), associated with the influence of organic matter, oxygen, and biomass of microorganisms, it takes the form

$$\rho = \rho_{\max}\frac{L_{ex}C_0}{L_{ex} - K_1C_0 + K_0L_{ex}} \cdot \frac{1}{1+\varphi a_i}, \quad (4.7)$$

where C_0 is the concentration of dissolved oxygen; K_1—a constant that characterizes the properties of organic substances; K_0—a constant that characterizes the effect of oxygen; *j*—coefficient of inhibition of the decomposition products of activated sludge; L_{ex}—is the concentration of the substrate in the medium; a_i—biomass of microorganisms.

In this regard, Eq. (4.7) is included in regulatory documents as one of the key calculation formulas that describe the biological nature of wastewater treatment processes.

In the oxidation of organic matter, part of the energy is dissipated, and part is transferred until all the organic matter carbon is oxidized to CO_2 and water. Therefore, the energy reserve of organic matter is depleted. Each substance has a certain energy reserve, i.e., it needs oxygen for complete oxidation. The amount of oxygen required for complete oxidation (BOD) measures the amount of organic matter that bacteria can oxidize under aerobic conditions. When assessing the level of decomposition of organic matter under anaerobic conditions and determining the efficiency of anaerobic structures where oxygen is not consumed, the BOD indicator should not be used.

The main difference between anaerobic fermentation and aerobic oxidation is that during the decomposition of organic matter under anaerobic conditions, the electron acceptor can be either bound oxygen of organic and inorganic compounds (anaerobic respiration), or intermediate reaction products inbound oxidation–reduction of some and the same substrate molecules (fermentation), not molecular oxygen. In both processes, the energy the cell receives during the decomposition of organic matter is stored in the high-energy bond of ATP. When one gram molecule of phosphate is cleaved from ATP, up to 42 kJ of energy is released, which is used by the cell in all metabolic reactions that require energy expenditure.

Anaerobic bacteria use the energy obtained during phosphorylation less efficiently than aerobic ones. This causes a much smaller increase in the biomass of microorganisms under anaerobic conditions compared to aerobic conditions, which contain the same amount of processed nutrients. However, anaerobic processes have advantages, such as disinfection of waste, biogas, and fertilizers.

The anaerobic process occurs in three stages: the stage of hydrolysis of polymeric compounds, the stage of acid fermentation (acidogenesis) and the stage of methanogenesis/sulfidogenesis/homoacetogenesis (terminal stage).

As a result of protein hydrolysis, polypeptides and amino acids are formed, which, eventually, when cleaved from them, amino groups are converted into fatty acids. Fats are broken down with the formation of glycerin and fatty acids. Carbohydrates under anaerobic conditions are also broken down into fatty acids [3].

Anaerobic conversion reactions of organic matter

At the terminal stage of anaerobic bioconversion, a number of biochemical reactions take place, revealing competition between homoacetoatogens, methanogens and sulfate reducers for the substrate.

Two pathways lead to sulfidogenesis or methanogenesis: the hydrogenotrophic (hydrogen) pathway and the acetoclastic (acetate) pathway. The remaining pathways are considered secondary in terms of importance. There is a bridge between the hydrogenotrophic and acetoclastic pathways, facilitated by a group of hydrogenotrophic homoacetogenic bacteria.

Methane formation by hydrogenotrophic methanogens:

$$4H_2 + CO_2 \rightarrow CH_4 + 2H_2O. \tag{4.8}$$

Formation of acetic acid by homoacetogens:

$$2CO_2 + 4H_2 \rightarrow CH_3COOH + 2H_2O. \tag{4.9}$$

Formation of hydrogen sulfide by hydrogenotrophic sulfate reducers:

$$5H_2 + SO_4^{2-} \rightarrow H_2S + 4H_2O. \tag{4.10}$$

Homoacetogens are also syntrophs for acetoclastic methanogens and acetotrophic sulfate reducers.

Methane formation by cytoclastic methanogens:

$$CH_3COOH + H_2O \rightarrow CH_4 + H_2CO_3. \tag{4.11}$$

Formation of hydrogen sulfide by sulfate reducers:

$$CH_3COOH + 2H^+ + SO_4^{2-} \rightarrow H_2S + 2CO_2 + 2H_2O. \tag{4.12}$$

Some bacteria are involved in several stages of decomposition: for example, they can use proteins and then carbohydrates, oxidize alcohols and then aldehydes, elemental and then bound nitrogen, etc. However, some bacteria can only consume certain substances without using others. Some types of bacteria can oxidize organic matter to the end—to the formation of carbon dioxide and water, and others—to intermediate products, which, in turn, can be a source for other types of bacteria. Therefore, in the treatment of wastewater containing a variety of organic and mineral substances, use only a mixed culture of bacteria, which has a wide range of physiological capabilities and resistance to external factors.

4.2 Thermal Processes of Aerobic and Anaerobic Purification of Semi-solid and Liquid Organic Waste

Interest in the problems of heat release during fermentation is primarily due to the need to remove it with cooling water, as well as the possibility of using the intensity of heat as an indicator of the metabolic activity of the biological agent, heating fermenters with their heat, etc. About 40–50% of total heat released during fermentation (for example, oxidation of organic waste in the biothermal process of composting) is accounted for by heat generated as a result of the biological agent's life and primarily due to energy metabolism. Accumulated high-energy compounds (mainly ATP) are consumed in biosynthesis reactions. For example, forming one ether bond in the molecule of any metabolite requires 12,570 J. In the reaction of ATP → → ADP + 50,280 J. Thus, one molecule of ATP releases so much energy that most of it (37,710 J) is converted into heat. This heat is released unevenly, with a maximum that falls during the maximum respiration rate and the highest rate of carbohydrate consumption during the period of active cell proliferation. Therefore, aerobic biofilters can be heated by their heat. It should be noted that a striking example of the use of the thermal effect of biological processes outside the fermenters is the process of composting organic waste. This is a biothermal process of decomposition of organic matter by aerobic microflora with facultative anaerobes to obtain organic fertilizer—compost. In the depth of the compost pile, the temperature can reach 70 °C [4].

The amount of heat generated by microorganisms is defined as

$$Y_{kcal} = \frac{Y_s}{H_s - Y_s \cdot H_c}, \tag{4.13}$$

where Y_{kcal}—expressed in g of cells/kcal of energy released; Y_S—g of cells/g of the consumed substrate; H_S and H_C—heat released during the combustion of the substrate and cells, respectively (determined experimentally), kcal/g.

The more oxidized the substance (glucose or acetate), the less heat is released per unit of biomass; the larger the value of μ, the more kcal. The rate of heat release per unit time (Q_w) is calculated based on the growth rate (μ), the volume of the bioreactor (V) and the coefficient Y_{kcal}:

$$Q_w = V \cdot \mu \cdot x \cdot \frac{1}{Y_{kcal}}. \tag{4.14}$$

Heat is also generated as a result of the mixer that mixes the culture fluid—"mixing heat":

$$Q_{mix} = N_r \cdot \tau_\varphi \cdot 3600\,J, \tag{4.15}$$

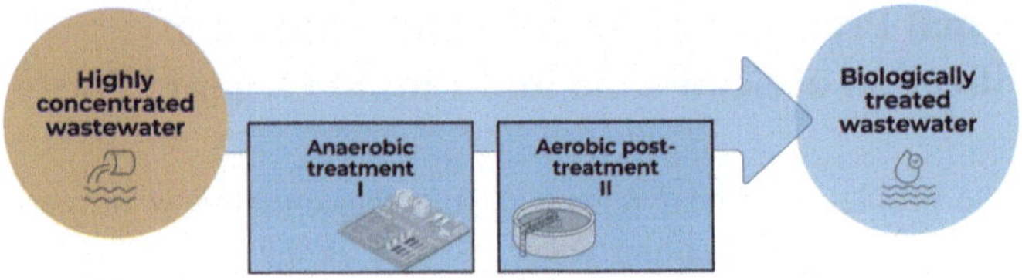

Fig. 4.2 Combined anaerobic–aerobic biological wastewater treatment

where N_r (W) is the power expended on mixing the gas–liquid mixture; τ_φ (hour)—fermentation time.

The thermal effect of mixing at the same number of revolutions of the will be higher as the viscosity of the culture medium increases. Therefore, as biomass accumulates, the thermal effect of mixing increases in the process of fermentation.

The third source of heat is *hot air* supplied to the fermenter for aerobic microorganisms; during bubbling, it is instantly cooled by the culture fluid and gives off the heat brought:

$$Q_{air} = V_{air} \cdot \rho_{air} \cdot C_{air} \cdot \left(t_{air} - t_f\right), \tag{4.16}$$

where $V_{air}, \rho_{air}, C_{air}, t_{air}$—respectively, the volume, density and temperature of the supplied air; t_f—the temperature at which the fermentation is carried out.

Usually, the thermal regime during fermentation is maintained at a certain level and is clearly controlled. All the generated heat must be removed by cooling water.

Regarding the thermal processes occurring in the fermenter, it can be stated that all of them (thermal effect of biosynthesis, mixing, heat brought by air, the heat of evaporation, heat transfer when cooled by water) take place simultaneously with different intensities and dynamism, hence the high design requirements heat exchangers.

Achieving the ultimate goals of biochemical treatment of organic waste and wastewater is based on the integrated use of anaerobic and aerobic processes (Fig. 4.2).

Their difference is that aerobic processes occur with the participation of strictly aerobic microorganisms in the presence of dissolved oxygen provided by energy-intensive aeration systems, and anaerobic processes are carried out strictly and optionally by anaerobic microorganisms in the absence of oxygen and, therefore, no need for energy-intensive systems [4].

In anaerobic processes, organic matter is not completely oxidized, and part of the energy of the original substrate is stored in a rather complex intermediate product of anaerobic fermentation and in the fermentation gases released. Residual intermediate fermentation products are capable of further oxidation with oxygen consumption, which makes it impossible to discharge anaerobically treated effluents directly into the reservoir because it can lead to oxygen deficiency. Thus, after anaerobic wastewater treatment, additional treatment is required. Such anaerobic–aerobic biological treatment is usually used for highly concentrated biodegradable organic contaminants in wastewater (e.g., food processing enterprises).

There is another important difference between aerobic and anaerobic processes, which is the amount of organic matter that is converted into cellular material, which characterizes the growth of biomass and the formation of precipitation that requires special treatment (compaction, stabilization, dehydration, etc.). This treatment complicates the technological scheme of treatment plants and is quite expensive and complicated. The degree of conversion of organic matter into biomass differs sharply for these two processes. Thus, in aerobic biological treatment with activated sludge, 50–80% of organic pollutants, expressed through TLV, are converted into biomass growth—excess activated sludge, which must be removed from the treatment system and specially treated, while in anaerobic treatment only 10% Organic pollutants are spent on biomass growth because most of the organic matter is consumed by anaerobic microorganisms to obtain energy, the yield of which in energy metabolism reactions during anaerobic fermentation is small, because organic matter is not completely oxidized. Part of the energy of the original substrate is stored in complex fermentation products. For these reasons, anaerobic fermentation processes are best for treating concentrated substrates of sewage sludge at sewage treatment plants.

The advantage of anaerobic processes over aerobic ones is determined by comparing their energy characteristics (Fig. 4.3).

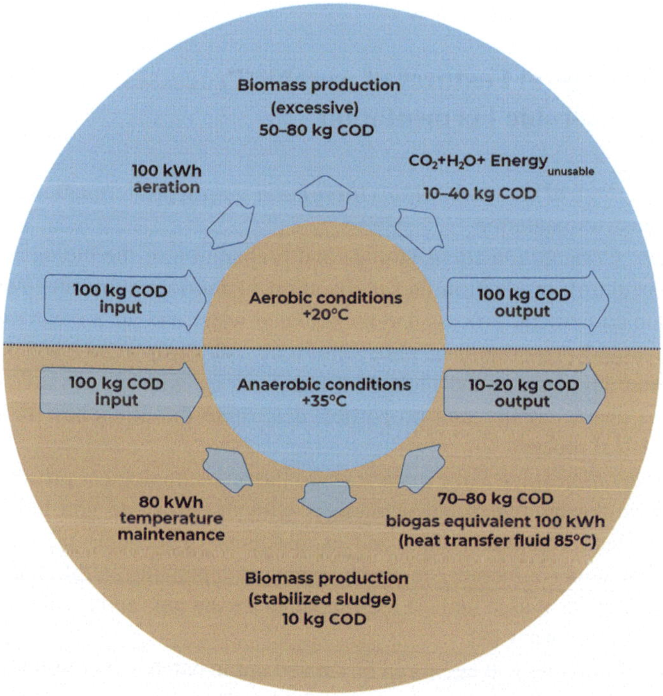

Fig. 4.3 Comparative energy characteristics of aerobic and anaerobic processes of biological wastewater treatment

As can be seen from Fig. 4.3, aerobic treatment consumes approximately 100 kW/h per 100 kg of COD contained in the source wastewater, which is used to provide aerobic processes with oxygen through the aeration system. In contrast, anaerobic treatment occurs with the release of biogas (combustible mixture of fermentation gases), equivalent to 190 kW/h, which can be converted into a coolant, part of which is spent on own needs, i.e. to maintain the thermal regime in anaerobic bioreactors; this amount of energy is approximately 80 kW/h. Thus, due to anaerobic biological treatment, the net energy yield is 190 – 80 = 110 kW/h per 100 kg of COD, in contrast to aerobic treatment, which consumes 100 kW/h per 100 kg of COD.

Anaerobically treated wastewater contains a residual amount of unoxidized fermentation products, which must be removed aerobically before discharging into the reservoir. These residual organic pollutants amount to 10–20 kg of COD/m^3 and may require not 10–20 kW/h for their oxidation, being more difficult to decompose, but twice as much—20 to 40 kW/h. Even in this case, combining anaerobic–aerobic processes for biological wastewater treatment can provide clean energy: 110 – 40 = 70 kW/h per 100 kg of COD by its production, utilizing biogas in cogeneration plants. This fact is very important for energy saving and increasing the reliability and efficiency of sewage treatment plants and environmental protection from the release of fermentation gases, which has a greenhouse effect.

4.3 Technological Features of Aerobic Processes and Anaerobic Fermentation

The biotechnological process contains three main stages: pre-fermentation, fermentation and post-fermentation.

At the pre-fermentation stage, storage and preparation of the biologicar agent's culture (inoculum), production and preparation of nutrient substrates and media, fermentation equipment, process and recirculated water and air are carried out [5]. Maintenance and preparation of pure culture are very important elements of the pre-fermentation stage, as the biological agent, along with its physiological and biochemical characteristics and properties, determine the efficiency of the entire biotechnological process.

The fermentation stage is the main stage in the biotechnological process because in its process, the biological agent interacts with the substrate and the formation of target products (biomass, endo- and exoproducts). This stage takes place in a biochemical reactor (fermenter). It can be organized depending on the characteristics of the biological agent used and the requirements for the type and quality of the final product in different ways [6].

Cultivation of biological agent can be carried out in batch and continuous modes, semi-continuously with substrate feeding, which will be discussed below [7].

The post-fermentation stage ensures the production of the final products and the disposal of waste and by-products. Depending on the localization of the final product (cell or culture fluid) and its nature at the post-fermentation stage, different equipment and methods of isolation and purification are used. The most time-consuming is isolating the product that accumulates in the cells.

According to the technological design, the following microbiological processes are distinguished: aerobic and anaerobic; solid-phase, surface and submerged; batch and continuous cultivation.

Aerobic cultivation—aeration of the medium—is a prerequisite in those microbiological processes that use aerobic biological agents (microorganisms).

The need for aerobic microorganisms in molecular oxygen depends on the oxidized carbon source and the microorganisms' physiological properties and growth activity. For the biosynthesis of 1 kg of yeast biomass, for example, 0.74–2.6 kg of molecular oxygen is required. With intensive substrate consumption, regardless of the carbon source, the biological agent assimilates 0.83–4.0 mgO_2/l of medium/min.

The solubility of oxygen in the medium is relatively low and depends on the temperature, pressure and concentration of dissolved, emulsified and dispersed components. At a pressure of 0.1 MPa and a temperature of 30 °C in 1 l of distilled water, the maximum amount of dissolved oxygen is 7.5 mg. However, when preparing the nutrient medium, the maximum solubility of oxygen varies in the range of 2–5 mg/l. Oxygen reserves in the environment ensure the viability of the aerobic biological agent for 0.5–2 min [8].

During biomass growth, microorganisms usually consume more oxygen than during the oversynthesis of the target metabolite. Sometimes, this indicates a critical concentration of oxygen, at which point cell respiration is limited. For most aerobic microorganisms growing in sugar-containing substrates, the critical oxygen concentration is 0.05–0.10 mg/l, corresponding to 3–8% of the total oxygen saturation of the medium. Limitation of growth and physiological activity of cells is observed at higher oxygen concentrations: on media with glucose, yeast growth is limited at pO_2 at 20–25% of full saturation.

The oxygen concentration of 50–60% of full saturation is considered optimal for biomass growth and 10–20% for biosynthesis of target metabolites.

Depending on what is the ultimate acceptor of hydrogen atoms or electrons, fermentation processes are divided into three groups: respiration (acceptor—oxygen), fermentation (acceptor—organic matter) and anaerobic respiration (acceptor—inorganic matter: nitrates, sulfates, etc.).

In ***obligate anaerobes***, fermentation is the only possible way to produce energy; in facultative anaerobes, it is a mandatory first stage of glucose catabolism, followed by aerobic oxidation of the resulting products if there is oxygen in the cultivation environment.

A separate intermediate group are aerotolerant microorganisms that receive the energy necessary for life in the anaerobic process, i.e. at the substrate phosphorylation level. At the same time, it has a respiratory chain that absorbs oxygen and creates favourable anaerobic conditions. This effect is called the "respiratory protection effect".

Anaerobic conditions in production are created by sealing the equipment and blowing the medium with inert gases, particularly gaseous products formed during fermentation [8]. The absence of the need for aeration of the cultivation environment somewhat simplifies the fermenter's (bioreactor) design during anaerobic fermentation and facilitates process control.

Surface cultivation on liquid substrates is conducted in a culture vessel containing the liquid medium and placed in chambers ventilated by air. The culture of microorganisms thus forms biomass in the form of a film or a solid layer on the surface of the liquid medium. The culture consumes oxygen directly from the gas phase—air. Mass transfer in such conditions is low.

Liquid nutrient media, even without stirring, avoided the disadvantages associated with agar impurities in the wash suspension and increased the yield of the process by using large containers for cultivation (bottles, fermenters). The use of liquid culture media with forced agitation of the culture for the cultivation of cells to equalize the growth conditions in different parts of the working volume of the cultivation tank has led to the emergence of dynamic systems of deep cultivation equipped with special equipment.

Submerged cultivation of microorganisms occurs in the entire volume of liquid nutrient medium containing dissolved substrate. The fermenter must ensure the growth and development of populations of microorganisms in the volume of the liquid phase, the supply of nutrients to the cells of microorganisms, the removal from microbial cells of their metabolic products (metabolism), and the removal of heat from the cells. In s*ubmerged cultivation* for aerobes, oxygen reserves in the nutrient medium are restored during the supply of aerated air—the oxygen absorption rate increases with the increasing intensity of mixing the medium.

Compared to surface cultivation, submerged cultivation in fermenters accelerates the growth and development of microorganisms by eliminating the "starvation zone" around the microbial cell and allowing for obtaining a homogeneous culture of "ideal mixing." The cells in such a system are evenly distributed over the volume of the fermenter and are always in the same conditions, i.e. nutrient substrates and metabolic products are also evenly distributed. "Ideal mixing" is essential for studying the processes of growth and development of the microbial population as a whole. The use of dynamic submerged cultivation systems has made it possible to rationally obtain not only biomass and endo-metabolites but also exoproducts of microbial synthesis released by the cell into the environment [9, 10].

Further improvement of the submerged cultivation process aims to develop control systems for cultivation conditions.

Batch method of cultivation

In the batch method of cultivation, the fermenter is filled with the initial nutrient medium and inoculum of microorganisms ($X_0 + S_0$ in Fig. 4.4). This method involves obtaining a culture to achieve a given phase of population development. The fermenter is filled with no more than 70% of its total volume. During a certain time in the apparatus, there is an interaction between the microorganisms and the substrate, which is accompanied by the formation of the product in culture ($X + S \rightarrow P$).

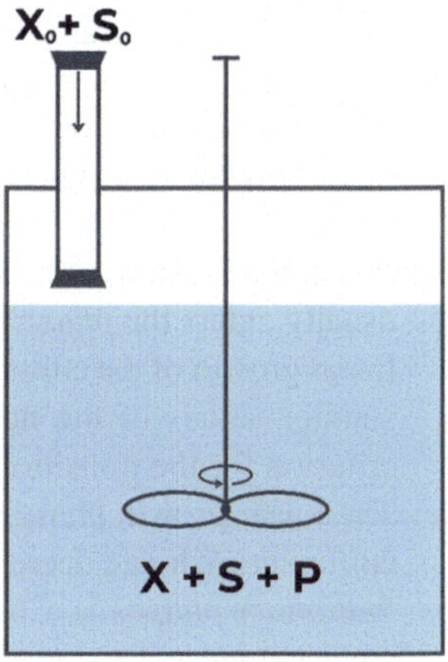

Fig. 4.4 Schematic of a batch bioreactor

When studying the growth dynamics of cultures of microorganisms, it is necessary to strictly adhere to particular conditions:

(1) Viability of seeding.
(2) The presence of all necessary nutrients in the cultivation medium.
(3) The absence in the medium of inhibitors that inhibit cell growth.
(4) Maintenance of optimal physical and chemical conditions in the medium.

Several growth phases can be distinguished in a simple homogeneous batch culture (Fig. 4.5).

1. *Lag phase (1).* When cells from a culture in a stationary phase are introduced into a fresh environment of the same composition, they acquire the ability to resume growth only after a certain period during which their chemical composition changes. Usually, during the transfer of cells from a nutrient-poor medium to a nutrient-rich one, the lag phase extends as the inoculum age increases.

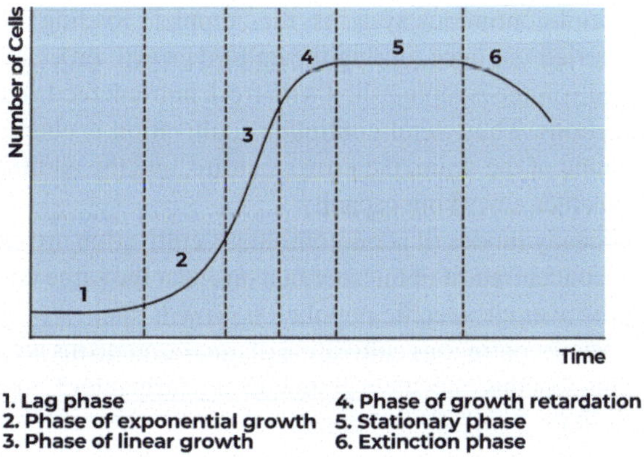

Fig. 4.5 The main phases of the growth curve

2. *Phase of exponential growth (2).* Exponential growth at high rates is usually not maintained in the microbial population for long. Typically, it is limited due to the depletion of available food sources, the accumulation of toxic metabolic products in the growth medium, or due to some changes in the physical properties of the medium. As a result, the growth rate decreases, and eventually, growth stops.
3. In a closed system, the exponential growth phase cannot develop indefinitely. It usually enters the *phase of linear growth (3),* which is characterized by uniform linear growth of the culture over time; there is a deviation from the points towards smaller values of the number of cells or products, which is an experimental criterion for the transition of the culture to the linear growth phase.
4. The linear growth phase may last for a relatively short period, during which the culture growth rate decreases to zero. This is *the phase of growth retardation (4).*
5. *Stationary phase (5).* Changes in the physical and chemical composition of the environment lead to the transition of the culture to a stationary phase, in which the increase in the number of cells does not occur. However, the cells still need energy sources to maintain their vital functions. The transition from the exponential phase to the stationary phase contains a period of unbalanced growth if different cellular components are synthesized at different rates.
6. *Extinction phase (6).* Several factors lead to cell death, the most important of which is the depletion of energy reserves in the cell. There is a sharp, exponential decrease in the number of viable cells. The rate of extinction depends on the species characteristics of the organism. For some purposes, a qualitative growth characteristic is sufficient, i.e. the usual observation of whether a crop is growing at all, but to have further, more complete information, it is necessary to measure growth quantitatively.

Batch cultures exist in an unstable state, which is a serious disadvantage, especially if they are used to study the properties of microorganisms. *Nearly all systems of batch cultivation are closed* because microorganisms in them multiply and pass all phases of development without inflow of nutrient medium and outflow of culture fluid [11].

Semi-continuous method of cultivation

In semi-continuous systems, the complete loading and unloading of the fermenter is carried out once, during the culture growth process, part of it is removed, and the freed volume is replenished with fresh nutrient feed. Thus, the drain-and-refill system functions. Thus, semi-continuous cultivation is characterized by the frequency and volume of the drain, the grown culture, and the addition of fresh nutrient feed to the fermenter's working capacity.

Steady modes of semi-continuous cultivation are characterized by fluctuations in the concentration of microorganisms near the same constant value and the constancy of the average specific population growth rate.

Semi-continuous cultivation of microorganisms is carried out in an open, dynamic, homogeneous one-stage system in any fermenter for batch cultivation equipped with a system of mixing and aeration.

Continuous cultivation method

In continuous cultivation processes, cells are constantly maintained in the exponential growth phase. To do this, fresh nutrient feed is continuously fed into the bioreactor, and the culture fluid's outflow containing the cells and the products of their vital activity is ensured. The main principle of continuous processes (as mentioned above) is a precise balance between the growth of biomass due to cell division and their reduction as a result of dilution with fresh feed. There are chemostatic and turbid-sufficient modes of continuous cultivation.

Continuous cultivation is carried out in an open dynamic system, which can be homogeneous and heterogeneous. This system is capable of long-term operation in a steady state and often uses homogeneous systems.

At continuous cultivation of microorganisms, it is necessary to prevent the leaching of culture from the system, i.e. to provide a constant concentration of cells. Under sterile conditions, a continuous, flowing method ensures the preservation of the culture in a physiologically active state for a long time.

Homogeneous systems in the fermenter can be single-stage, two-stage and multi-stage. Continuous cultivation in one bioreactor is called single-stage. Multi-stage cultivation involves sequential or cascading placement of bioreactors, which allows it to introduce the principle of differentiated modes in continuous biotechnological processes based on creating a system of bioreactors [12].

Complete-mixing cultivation systems. Theory of chemostatic cultivation

In *the chemostatic mode* of cultivation, the self-regulating system arises from the following reasons: if the initial intake of fresh nutrient medium and leaching of biomass exceeds the rate of cell division, then as a result of breeding the culture decreases the concentration of substances that limit growth processes, and the growth rate of the culture increases; the growing population begins to more actively consume the substrate, which, in turn, leads to inhibition of crop growth. The end result of these processes is (after a series of damped oscillations) to establish a balance between the growth rate of the culture and its breeding [13].

For chemostatic culture, there are certain quantitative relationships between biomass concentration x, substrate concentration s and growth rate μ. The growth of the culture is characterized by a chemostatic curve depicting the dependence of the concentration of cells x and the concentration of substrate s on the dilution rate D (Fig. 4.7).

By changing the concentration of the factor that limits growth, you can change the population density. By varying the dilution rate, regimes can be obtained that ensure different population growth rates. During the slow flow of the medium, i.e. with slow growth, the culture experiences severe limitation and deep starvation for this substrate. During the rapid flow of the medium, i.e. with rapid growth, the level of starvation is low, approaching the conditions of exponential growth [14].

For all organisms there is the same probability of removal from the reactor, and the rule of cross-curves is fully realized. Stationary concentrations of substrate, product and active biomass are established (Fig. 4.6).

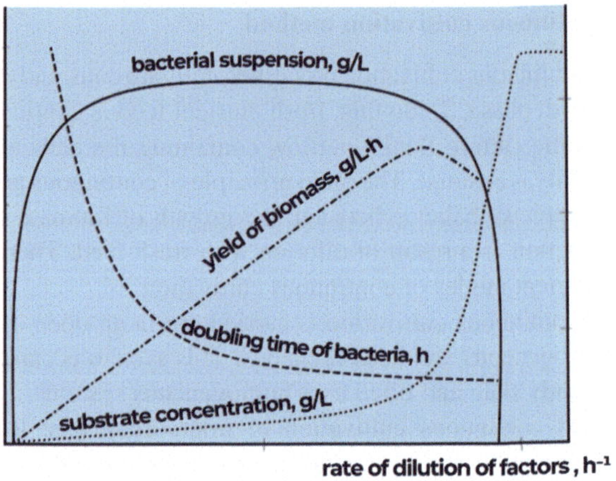

Fig. 4.6 Chemostatic curve: The relationship between the density of the bacterial suspension (g/l), substrate concentration (g/l), biomass yield (g/l h), doubling time of bacteria (h), and dilution rate (h^{-1}) factors that limit the effect on pure cultures, and competitive relationships—on mixed

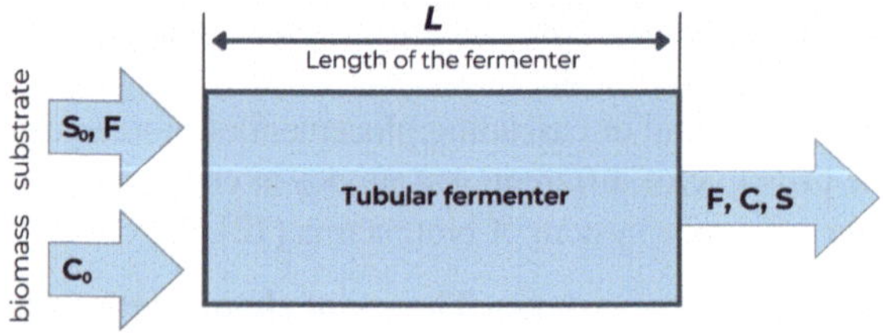

Fig. 4.7 Scheme of the tubular continuous process of cultivation of microorganisms: S_0—the concentration of the substrate in the incoming medium; S is the concentration of the substrate in the flowing medium; C_0—initial concentration of biomass; C—leakage biomass concentration; F is the volume flow rate of the liquid; L is the length of the tubular fermenter

The dilution rate at a stable steady state, if the culture develops at a constant rate and all conditions are unchanged, is equal to the specific growth rate:

$$D = \frac{1}{x} \cdot \frac{dx}{dt} = \mu. \tag{4.17}$$

The specific growth rate is the increase in biomass per unit of time.

The dilution rate of the medium in the fermenter per unit volume D is equal to the ratio of the flow rate to V—the volume of the culture. The inverse *of* $V/F = t_r$ is the time of replacement of the existing medium with a new one or the time spent in the fermenter of each medium element.

The biomass balance x present in the fermenter during the steady state is as follows: change in the amount of biomass = growth—biomass lost through outflow.

$$\text{Growth rate} = \frac{dx}{dt} = (\mu - D) \cdot x. \tag{4.18}$$

Substrate balance that limits growth: change in the amount of substrate = inflow—attribution of substrate consumption by microorganisms.

$$\text{Substrate consumption rate} = \frac{ds}{dt} = D \cdot (s_r - s) \cdot \frac{\mu \cdot x}{Y}, \tag{4.19}$$

where Y is the economic coefficient or the proportion of consumed substrate spent on biomass synthesis; s_r is the concentration of the substrate in the incoming medium; s is the concentration of the substrate in the fermenter.

The rate of growth in the steady state in the fermenter

$$\mu = \frac{\mu_m \cdot s}{(s + K_s)}, \tag{4.20}$$

where μ_m is the maximum possible growth rate; K_s is the Monod constant. It is numerically equal to the residual concentration of the substrate, which, limiting growth, slows it twice.

The steady-state concentration of substrate s in the fermenter is determined by the rate of dilution and kinetic parameters of the growth of the microorganism. It does not depend on the initial concentration of the substrate:

$$s = \frac{K_s \cdot D}{(\mu_m - D)}. \tag{4.21}$$

The steady-state concentration of biomass linearly depends on the concentration of the substrate solution introduced into the fermenter:

$$x = Y \cdot (s_r - s) = Y \cdot \left[\frac{s_r - K_s \cdot D}{(\mu_m - D)} \right]. \tag{4.22}$$

In the chemostat, it is practically possible only to approach μ_m, but not to reach it. The dilution rate at which biomass is leached from the fermenter is called D critical (cr). *The growth of the culture in the chemostat mode is characterized by a critical dilution rate D_{cr}, above which there is no growth of the culture.* Theoretically, $D_{cr} = \mu_m$, but since the concentration of the substrate, which limits the apparatus, can not be greater than s_r, then

$$D_{cr} = \mu_m \cdot \frac{s_r}{s_r + K_s}. \tag{4.23}$$

Fermenter productivity from biomass yield

$$R = D \cdot x. \tag{4.24}$$

Knowing the basic constants K_s, μ_m and Y, you can calculate the culture parameters x and s for any value of $\mu(s) = D$.

The steady-state concentration of the product is linearly related to the concentration of the introduced substrate:

$$P_{st} = \frac{M_{st}}{Y_P}, \quad (4.25)$$

$$P_{st} = \frac{Y_S}{Y_P}\left(s_r - \frac{D \cdot K_S}{\mu_m - D}\right) \quad (4.26)$$

The dependence of the steady-state concentration of the product on s_r and D is similar to the dependence of these variables on the steady-state concentration of the cell [15, 16].

Full displacement cultivation systems

This method of cultivation is used for anaerobic conditions. The open system of full displacement differs from the system of ideal mixing in that the culture in it is not mixed and is a flow of liquid through the tube. The most common aparatus is a tubular reactor (Fig. 4.7). It can have various forms (straight, S-shaped, spiral) and is established horizontally or vertically. The complete displacement system represents a spatial, continuous-flow variation of batch culture. This culture, during the time from sowing to unloading, goes through all stages of batch culture, i.e. the growth phases are distributed not in time but in space, and each part of the fermenter in a constant mode corresponds to a certain segment of the growth curve. Sowing is carried out continuously at the entrance to the fermenter simultaneously with the supply of medium. According to this principle, the fermentation stage is carried out during the production of beer in tower continuous-flow tanks. It should be noted that at this time, there are fermentation apparatuses that provide processes with a regime approaching complete displacement and aerobic cultivation. These are rotating tubular reactors with packing or internal aerating elements, as well as multi-section column apparatus.

During this process, the nutrient medium and inoculum are continuously fed into the apparatus, without back-mixing. The device is made in the form of a long tube of large diameter [17].

4.4 Solid Phase and Liquid Phase Fermentation—Application in Biotechnological Processes of Environmental Protection

The traditional application of solid phase fermentation (SPF) is the production of secondary metabolic products (antibiotics, alkaloids, growth hormones, etc.), biofuels, enzymes, organic acids, aromatic compounds, biological detoxification of agro-industrial waste, food additives, biopharmaceuticals and others. This technology has attracted attention as an alternative to liquid-phase fermentation (LPF) in environmental biotechnology (Table 4.1). The products obtained by SPF are more stable, energy consumption costs are lower, the volume of fermenters is smaller, and the volumes of polluting effluents are significantly reduced.

Solid-phase fermentation is usually carried out in a solid, granular, or viscous medium, the humidity of which is 30–80%.

There are three types of solid-phase processes [18]:

- surface processes: a layer of substrate, such as straw, does not exceed 3–7 cm ("thin layer"); large aluminium trays, up to several square meters in size, or cultivation chambers function as bioreactors).
- deep solid-phase processes in an unmixed layer ("high layer"): bioreactors are deep open vessels. Devices for diffusive and conjunctive gas exchange have been developed for aerobic processes.
- solid-phase processes in a mixed and aerated substrate mass, which may be homogeneous or consist of solid substrate particles suspended in the liquid.

If the substrate is granular, then some solid particles are in good contact with air; the growth of microorganisms, in this case, occurs mainly on the surface of solid particles, as well as in pores filled with either water or air [19–21].

A comparison between LPF and SPF revealed the following:

1. Low water content reduces the possibility of contamination by bacteria and yeast.
2. Environmental conditions are similar to the natural habitat for fungi, which are the main group of microorganisms used for SPF.
3. A higher level of aeration is especially necessary in processes with intensive oxidative metabolism.
4. Inoculation, together with spores, facilitates uniform scattering in the environment.
5. Solid substrates usually provide all the nutrients needed for colony growth.
6. Enriched substrates allow simpler and more economical designs of bioreactors.
7. Low energy consumption—in some cases there is no need for autoclaving, steam treatment, mechanical mixing and aeration.
8. Due to the high concentration of the product, the need for special solvents is reduced.

Table 4.1 The use of solid phase fermentation in various sectors of the economy

Sector	Example
Agro-industrial industry	Biotransformation of agricultural waste Food additives
Environmental monitoring	Bioremediation and processing of hazardous components Biological detoxification of agro-industrial waste Production of biopesticides
Industrial fermentation, green energy	Production of enzymes, biologically active substances, organic acids, biofuels

9. Low humidity can have a positive effect on the production of certain products that can not be cultivated in the LPF.
10. Productivity for products obtained by SPF is higher.

At the same time, SPF has a number of disadvantages compared to LPF:

1. Only microorganisms capable of growing at low humidity levels may be used.
2. Substrates usually require pre-treatment (size reduction, homogenization, physical, chemical or enzymatic hydrolysis, boiling or steam treatment).
3. Complicated analysis of biomass parameters (humidity, pH, mass transfer, heat transfer).
4. There are difficulties in controlling the process parameters (pH factor, moisture content and concentration of substrate, oxygen and biomass).
5. Many important scientific and technical aspects are poorly understood. Information on the design and operation of large-scale reactors is insufficient.
6. The possibility of contamination with unwanted fungi.
7. Difficulty in diverting the metabolic temperature produced during growth.
8. Extracts containing products obtained from SPF are often viscous.
9. The spread of the colony limits mass transfer.
10. The increase of the substrate layer complicates the supply of microorganisms with oxygen.
11. Cultivation time is increased due to the need for germination spores.
12. Cultivation time is longer than in the LPF.
13. Stirring of the layer is not allowed if mycelial organisms, such as micromycetes, are cultivated, and due to the lack of mixing, the growth of microorganisms occurs on the principle of colonization, so there is often a local lack of nutrients.

A significant part of the identified shortcomings of SPF compared to LPF can be removed when using the results of an in-depth study of solid-phase fermentation. A number of researchers claim that SPF is more efficient and economical than LPF in the production of a wide range of organic products (feed, enzymes, organic acids, flavourings, antibiotics, compost, biopesticides, etc.). When choosing microorganisms for cultivation, it is necessary to take into account their specific properties.

Bacteria and yeast can also be cultured in a solid phase. However, mycelial fungi are best adapted to SPF due to their physiological, enzymological and biochemical properties. The filamentous growth of fungi allows mycelial fungi to enter solid substrates, which also gives them a major advantage over unicellular microorganisms for colonising the substrate and using available nutrients. In addition, their ability to grow in conditions of low water activity and high osmotic pressure makes fungi efficient and competitive in the natural microflora for the bioconversion of solid substrates.

The main factors influencing the results of SPF are the properties of the substrate, such as particle size, shape and porosity, as well as mass transfer, temperature, acidity and moisture content of the medium.

Various authors have widely described and reviewed the role of substrate moisture content. Moisture content is a critical factor in the SPF process that affects the growth and biosynthesis of metabolites. The optimal level of substrate humidity varies between 30 and 75%. Lower moisture content causes a decrease in nutrient solubility and sporulation. Higher humidity levels cause steric hindrance that impede growth, reduced porosity and oxygen exchange, and increased bacterial contamination risk.

There are two types of mass transfer in SPF: on a microscale (transfer to and from cells of a microorganism) and on a macroscale (oxygen transfer, convection) outside cells.

An increase in temperature in SPF results from metabolic activity if the removal of high temperature is insufficient. This directly affects the germination of spores, growth and product formation. Achieved temperature level—a function of the type of microorganism, porosity, particle diameter and depth of the base.

Temperature control in SPF is somewhat more difficult than in LPF due to the nature of the substrate. Thus, the control methods used in LPF do not apply to SPF. The traditional method of control is ventilation. High aeration levels can reduce water activity, so saturated oxygen is used. The pH level is controlled by adding sources of nitrogen (ammonium and nitrate salts) [21].

Solid–liquid systems include multiphase systems in which the culture grows at the boundary of different phases: liquid–solid phase, liquid–solid phase—gas. In these systems, the cells are held by adhesion to a solid base—the filler and multiply on it, forming a biomass film. A typical example is the production of vinegar in trickling reactors.

In aerobic growth processes, the limiting factors are probably oxygen and substrate. In thin films of biomass, each of the microbial cells attached to the surface is fully supplied with a nutrient medium and can grow and multiply at a maximum exponential rate. As cells form a thicker biomass film, their growth is limited by the diffusion of substrate and oxygen into the film.

The cultivation of microorganisms that form films from biomass is carried out in a packed-column fermenter. A macrocarrier (coke, wood chips, sawdust, glass beads, etc.) or a microcarrier (amberlite resins, Sephadex particles, etc.) can be used as a filler. Cells grown using this method are known as immobilized cells.

In industrial microbiology, solid–liquid systems have been used in "irrigation filters" in wastewater treatment, in producing organic solvents and acids, in the fermentation of wood hydrolysate to alcohol, and so on.

Solid culture is also used to study the metabolic activity of natural microbial populations. When passing various solutions through a column of soil sample, which is a carrier, models of conditions close to natural are obtained. Multiphase systems in beer production have become widespread.

Multiphase systems of the solid–liquid type include those in which the solid phase consists of the culture itself, forming films on the surface of the flowing liquid nutrient medium. The growth of microorganisms, in this case, is carried out in a special horizontal vessel. Nevertheless, the use of liquid-phase cultivation dominates in modern environmental biotechnology. Although it is necessary to take into account the specificity of existing areas of biotechnological processes (bio-wastewater treatment, biodegradation of oil pollution, bioremediation of contaminated soils, etc.)

and, accordingly, the possibility of using solid-phase and liquid-phase fermentation for their implementation [22–24].

References

1. Encyclopedia of Ecology (2008) Ammonification. Earth Planet Sci. https://www.sciencedirect.com/topics/earth-and-planetary-sciences/ammonification
2. ScienceDirect (2024) Nitrobacter. Biochem Gen Mol Biol. https://www.sciencedirect.com/topics/biochemistry-genetics-and-molecular-biology/nitrobacter
3. Adekunle K, Okolie J (2015) A review of biochemical process of anaerobic digestion. Adv Biosci Biotechnol 6:205–212. https://doi.org/10.4236/abb.2015.63020
4. Aziz A, Basheer F, Sengar A, Irfanullah I, Khan SU, Farooqi IH (2019) Biological wastewater treatment (anaerobic-aerobic) technologies for safe discharge of treated slaughterhouse and meat processing wastewater. Sci Total Environ 686:681–708. https://doi.org/10.1016/j.scitotenv.2019.05.295
5. Albernas-Carvajal Y, Corsano G, Kafarov VV, Cortés MG, Suárez EG (2014) Optimal design of pre-fermentation and fermentation stages applying nonlinear programming. Energy Convers Manage 87:1195–1201. https://doi.org/10.1016/j.enconman.2014.04.011
6. Magar ST, Aryal S (2024) Bioreactor: design, principle, parts, types, uses, diagram. Microbe Notes. https://microbenotes.com/bioreactor-design-principle-parts-types-uses-diagram/
7. Barta DG, Cornea-Cipçigan M, Margaoan R, Vodnar DC (2022) Biotechnological processes simulating the natural fermentation process of bee bread and therapeutic properties—an overview. Front Nutr 9:871896. https://doi.org/10.3389/fnut.2022.871896
8. Trends in Food Science & Technology (2018) Aerobic conditions. Agric Biol Sci. https://www.sciencedirect.com/topics/agricultural-and-biological-sciences/aerobic-conditions
9. Steward K (2024) An introduction to culturing bacteria. Technol Netw Immunol Microbiol. https://www.technologynetworks.com/immunology/articles/an-introduction-to-culturing-bacteria-355566
10. Mészáros É (2022) How to culture bacteria. INTEGRA Biosci Blog. https://www.integra-biosciences.com/global/en/blog/article/how-culture-bacteria
11. Kumakura D, Yamaguchi R, Hara A, Nakaoka S (2023) Disentangling the growth curve of microbial culture. J Theor Biol 573:111597. https://doi.org/10.1016/j.jtbi.2023.111597
12. Encyclopedia of Microbiology (2019) Continuous culture. In: Encyclopedia of microbiology, 4th edn. https://www.sciencedirect.com/topics/immunology-and-microbiology/continuous-culture.
13. Drake DR, Brogden KA (2002) Continuous-culture chemostat systems and flowcells as methods to investigate microbial interactions. Polymicrob Dis. https://www.ncbi.nlm.nih.gov/books/NBK2488/
14. Allman T (2020) Continuous culture for beginners. Infors HT Blog. https://infors-ht.com/en/blog/continuous-culture-for-beginners
15. Ellermeyer S, Hendrix J, Ghoochan N (2003) A theoretical and empirical investigation of delayed growth response in the continuous culture of bacteria. J Theor Biol 222(4):485–494. https://doi.org/10.1016/S0022-5193(03)00063-8
16. Herbert D, Elsworth R, Telling RC (1956) The continuous culture of bacteria; a theoretical and experimental study. J Gen Microbiol 14:601–622
17. Jaibiba P, Vignesh SN, Hariharan S (2020) Working principle of typical bioreactors. In: Bioreactors. INC. https://doi.org/10.1016/B978-0-12-821264-6.00010-3
18. Biotechnology of Microbial Enzymes (2017) Solid state fermentation. In Immunology and microbiology. https://www.sciencedirect.com/topics/immunology-and-microbiology/solid-state-fermentation

19. Zhao Y-S, Eweys AS, Zhang J-Y, Xiao X et al (2021) Fermentation affects the antioxidant activity of plant-based food material through the release and production of bioactive components. Antioxidants 10(12):2004. https://doi.org/10.3390/antiox10122004
20. Lee SY, Ra CH (2022) Comparison of liquid and solid-state fermentation processes for the production of enzymes and beta-glucan from hulled barley. J Microbiol Biotechnol 32(3):317–323. https://doi.org/10.4014/jmb.2111.11002
21. Mattedi A, Sabbi E, Farda B, Djebaili R, Mitra D, Ercole C et al (2023) Solid-state fermentation: applications and future perspectives for biostimulant and biopesticides production. Microorganisms 11(6):1408. https://doi.org/10.3390/microorganisms11061408
22. Wojtowicz K, Steliga T, Kapusta P, Brzeszcz J (2023) Oil-contaminated soil remediation with biodegradation by autochthonous microorganisms and phytoremediation by maize (*Zea mays*). Molecules 28:6104. https://doi.org/10.3390/molecules28166104
23. Sharma I (2020) Bioremediation techniques for polluted environment: concept, advantages, limitations, and prospects. IntechOpen. https://doi.org/10.5772/intechopen.90453
24. Romantschuk M, Lahti-Leikas K, Kontro M, Galitskaya P, Talvenmäki H, Simpanen S et al (2023) Bioremediation of contaminated soil and groundwater by in situ biostimulation. Front Microbiol 14:1258148. https://doi.org/10.3389/fmicb.2023.1258148

Chapter 5
Engineering Implementation of Environmental Biotechnology

Abstract Engineering realization of processes of disinfection and treatment of wastewater and its sludge, as well as waste treatment in biotechnological processes, has considerable variability. Therefore, in this section, the basic technical methods and schemes of realization of such processes and an introduction to such biotreatment are given. Examples of biological wastewater treatment realisation in aerobic and anaerobic conditions are considered. Design solutions of biogas plants for recycling various organic wastes and their prevalence in different countries are considered.

5.1 Equipment for Biotechnological Productions

Bioengineering deals with the technical support of biotechnological processes. There is a wide variety of equipment for the fermentation process and for isolating and obtaining the finished product [1].

The most complex and specific equipment for the fermentation stage. Technically, the most complex fermentation processes are aerobic, submerged, sterile and continuous (or with substrate feeding). Apparatus for surface and anaerobic fermentation are less complex and energy-intensive. The modern literature describes hundreds of bioreactors that differ in design, principle of operation and size (from several liters to several thousand cubic meters). The number of cultivation methods and the extraordinary variety of biological agents used have led to a vast variety of design solutions, depending on a number of factors: the type of biological agent and media, technology and scale of production, as well as the target product and others. The technical equipment of biotechnology is based on the general provisions of technical biochemistry and food technology, but it has its specifics. The fundamental difference between biotechnological processes and purely chemical ones is as follows:

- Sensitivity of biological agents to physical and mechanical effects.
- The presence of interphase transfer of substances (type "liquid-cells", "gas–liquid-cells").
- Requirements of asepsis conditions.
- Low speeds of many processes in general.

Y. Chernysh et al., *Sustainable Biotechnology and Environmental Protection*, Green Energy and Technology, https://doi.org/10.1007/978-3-031-89436-7_5

- Instability of target products.
- Foaming.
- Complexity of mechanisms of growth regulation and biosynthesis.

Depending on the process's peculiarities, various fermenters or bioreactors are used in microbiological productions.

Bioreactors are usually sealed cylindrical tanks with a height of 2–2.5 times the diameter. Most often, they are made of stainless steel. A double casing or coiled heat exchanger is used to maintain the reactor's temperature. The working volume of the bioreactor usually does not exceed 7/10 of the total volume.

According to the structure of the streams, bioreactors can be devices for complete mixing or complete displacement.

Structural differences of bioreactors are determined mainly by the methods of energy supply and mixing:

- Bioreactors with energy supply to the gas phase.
- Bioreactors with energy supply to the liquid phase.
- Bioreactors with combined energy supply.

The gas mixtures used in such systems may be of different compositions depending on the type of cultured biological objects. For example, in the growth of anaerobic chemolitotrophic microorganisms, a mixture of carbon dioxide and hydrogen of a certain ratio is used.

There is also a classification of bioreactors by the method of mixing, according to which device is used, either mechanical, pneumatic or circulating mixing [2].

5.2 Aeration Systems for Wastewater Treatment

5.2.1 Classification of Aeration Tanks

Aeration tank belongs to homogeneous bioreactors. A typical bioreactor design is a reinforced concrete sealed vessel of rectangular cross-section connected to a settling tank. The aeration tank is divided by longitudinal partitions into several corridors, usually 3–4. Structural differences between different types of aeration tanks are mainly related to the configuration of the bioreactor, the oxygen supply method, and the load amount. Typical schemes for aeration tanks are given in Fig. 5.1. The process of biocleaning in the aeration tank consists of two stages. The first stage is the interaction of the settled wastewater, containing about 150–200 mg/l of suspended particles and up to 200–300 mg/l of organic matter, with air and particles of activated sludge in the aeration tank for some time (from 4 to 24 h and higher depending on the type of runoff, requirements for the depth of treatment, etc.). On the second, water and activated sludge particles are separated in the secondary settling tank. Biochemical oxidation of organic substances of effluents in the aeration tank at the first stage is realized in two stages: at the first, microorganisms of activated sludge adsorb

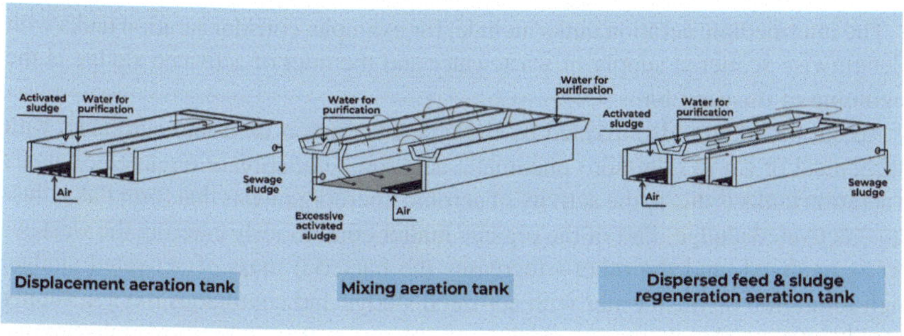

Fig. 5.1 Schemes of aeration tanks

pollutants of effluents, and at the second—oxidize them and restore the oxidizing ability.

The designs of the used aeration tanks are divided according to the method of wastewater supply and their flow into three main types (Fig. 5.1): displacement aeration tanks with a "piston-like" wastewater flow, mixing aeration tanks with dispersed or central supply and discharge, and intermediate-type aeration tanks combining features of both.

Displacement aeration tanks include single-, double-corridor, etc., aeration tanks, in which the corridors are separated from each other by longitudinal guide partitions that do not extend to one of the end walls. At the ends of the aeration tank are channels for inlet and outlet of wastewater. Depending on the geometric dimensions, these aeration tanks meet the condition of complete wastewater displacement to varying extents. A feature of the process that takes place in aeration tanks-displacers is the change in the concentration of pollutants in wastewater and the rate of treatment along the length of the aeration tank. The oxidation process in the aeration tanks-displacers is uneven: at the beginning of the aeration tank—it is faster, and as it approach the end and reduce the amount of substrate—it is slower [3].

Displacement aeration tanks are more suitable in treating wastewater of complex composition, which contain a significant proportion of industrial discharges with pollutant with a concentration of pollutants on the BOD not more than 500 mg/dm^3.

In mixing aeration tanks, the incoming wastewater portions are almost instantly mixed with the whole mass of treated wastewater and activated sludge, which allows to evenly distribute oxygen and the load in organic matter on the activated sludge.

They provide complete and fast sewage mixing with a mass of activated sludge; in a steady mode, they work with uniform speeds in the clearing process. The predominant use of aeration tanks-mixers in the treatment of highly concentrated industrial wastewater, similar in composition to domestic (food processing plants, breweries and fish factories), as well as uneven inflow and volley overloads, which often occur, was justified by Professor Bazyakina in 1948. However, when using mixers, there is a risk of bulking of activated sludge; in any case, they are more prone to it than other structures of aeration tanks due to high loads on activated sludge in the entire volume of structures.

The intermediate aeration tanks include, for example, corridor aeration tanks with a lengthwise scattered supply of wastewater and the inlet of activated sludge at the beginning of the corridor.

Aeration tank are also divided by the type of aeration used into aerotanks with mechanical or (most common) pneumatic aeration. Oxidation of organic pollutants in aeration tanks is due to the activity of aerobic microorganisms that form flaky clusters—activated sludge. Part of the organic matter continuously entering the wastewater is oxidized, and the other—increases the bacterial mass of activated sludge. High saturation of wastewater with activated sludge and continuous oxygen supply provide intensive biochemical oxidation of organic matter, so aeration tanks are one of the most advanced facilities for biochemical treatment. Depending on the required degree of reduction of the content of organic pollutants, aeration tanks are designed for complete biological (content in purified waters of BOD_5—20 to 25 mg/dm^3, NO_3—not less than 5.0–6.0 mg/dm^3) and partial ($BOD_n > 25$ mg/dm^3) purification.

Widely known in the practice of water treatment corridor aeration tanks with filter channels (Fig. 5.2) are difficult to operate and are bulky. In addition, the sensitivity of this facility to overload significantly limits its applicability to industrial wastewater.

The main factor that divides aeration tanks into high- and low-load is the oxidizing capacity, i.e. the amount of BOD (biochemical oxygen demand) in wastewater removed from 1 m^3 aero tank for 1 day. Usually, the load on the sludge with incomplete biological treatment is from 500 to 2000 mg/g per 1 day; with complete biological wastewater treatment, the load on the sludge varies from 200 to 500; with a smaller value, the nitrification process is possible. In high-load aeration tanks, the treatment process takes 0.5–2 h, as a result of which the hydraulic load is more than 20 m^3/day per 1 m^3 of the structure and the daily load on the sludge for BOD_n – 0.9 kg/kg with an efficiency of 70–95%. Highly loaded sludge comprises 80–85% of organic matter, and mineralized sludge comprises 60–65%.

In practice, the operation of aeration tanks uses two characteristics of the sedimentation properties of activated sludge—the dose of sludge by volume and sludge index. The sludge dose by volume characterizes the ability of activated sludge to precipitate within 30 min of settling.

Approximately the dose of sludge can be taken according to Table 5.1.

Relatively high doses (6–8 g/dm^3) can be maintained in settling tanks. The design feature of this structure is that the aeration tank is integrated with a secondary settling

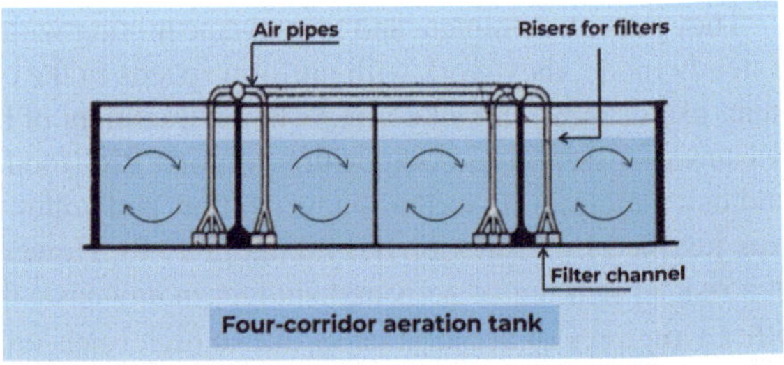

Fig. 5.2 Four-corridor aeration tank

Table 5.1 Approximate values of the sludge dose depending on the type of aeration tank and the concentration of pollutants (according to BOD_n)

BOD_n wastewater, mg_{O2}/dm^3	Sludge dose, g/dm^3, depends on the type of aeration tank		
	aeration tank without a regenerator	settling aeration tanks	aeration tank with a regenerator
Up to 100	1.2	3	–
100–150	1.5	3.4	–
150–200	1.8	3.7	$\alpha_{air} \approx 1.5$ g/l
More than 200	1.8–3	4–5	$\alpha_{reg} \approx 4$ g/l

tank (Fig. 5.3). The disadvantage of many existing designs of aeration tanks-settlers is the instability of the suspended layer of sludge, filtered in the settling zone, which depends on fluctuations in flow and composition of wastewater.

The sludge index is the volume occupied by 1 g of dry sludge after 30 min of settling in a 1 dm^3 cylinder. The sludge index also characterizes the sedimentation properties of sludge, but taking into account its dry mass.

When estimating the sludge dose by volume, it is required to note the measurement results every 3 min of settling to determine the smoothness and uniformity of deposition, which is a characteristic of good sedimentation properties of sludge, as well as its ability to occupy the smallest volume after 30 min of settling.

The sludge index I (cm^3/g) is calculated after the dry weight and volume dose values have been obtained. The result is obtained by dividing the values of the dose of sludge by volume V, cm^3/dm^3, by quantitative values of the dose of sludge by dry weight d, g/dm^3:

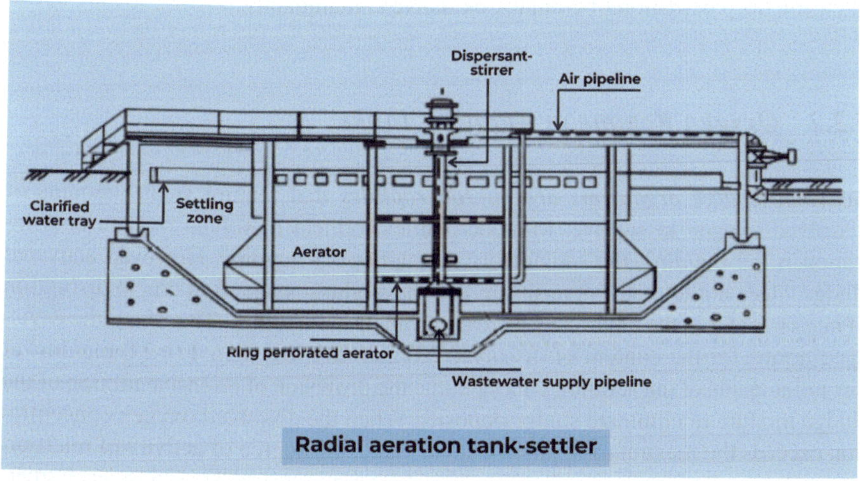

Fig. 5.3 Aeration tank settler radial

$$I = \frac{V}{d}. \quad (5.1)$$

Activated sludge is deposited quickly with the formation of a light zone (deposition rate of 1 m in 1 h or more), smoothly, well compacted, not occupying a large volume after the final compaction and not floating after deposition for 1.0–1.5 h.

Depending on the technical capabilities for the timely removal of settled sludge from secondary settling tanks, each specific biological treatment facility will have its own optimal sludge index values.

For treatment plants with aeration tanks, the optimal sludge index values are conventionally considered to be between 80 and 120 cm^3/g, with an acceptable range of 60–150 cm^3/g.

One of the main requirements for the sludge index is the stability of its values, which indicates satisfactory living conditions of sludge and satisfactory operation of facilities (the optimal amount of sludge is removed from the system, and a normal dose of return sludge is maintained).

The time of contact of activated sludge with polluted wastewater is determined by such a technological parameter as the aeration period, calculated by the formula

$$T = \frac{W}{q}, \quad (5.2)$$

where W is the volume of aeration facilities, m^3;q—hourly flow rate of wastewater, m^3/h.

The duration of the aeration period is assumed during the design and is determined by the complexity of the composition of treated industrial wastewater. The more complex this composition, the longer contact of wastewater with sludge is required to ensure deep oxidation of complex oxidative contaminants.

5.2.2 *Oxygen Regime in Aeration Tanks*

Activated sludge organisms are microaerophiles that require small amounts of dissolved oxygen to survive. Microaerophiles' critical dissolved oxygen concentration is 0.2 mg/dm^3; the satisfactory amount is 0.5 mg/dm^3. However, activated sludge "does not tolerate" deposits and, at the slightest stagnation due to disruption of mass transfer in the flakes, begins to die from its metabolites. Therefore, the envisaged norms for the content of dissolved oxygen (not less than 1.0–2.0 mg/dm^3 at any point inside of the aeration tank) assume the provision of intensive mixing of the sludge mixture to eliminate sludge deposits. When the dissolved oxygen concentration exceeds the maximum required critical value, the degree of activity of microorganisms does not increase, and purification does not improve. Therefore, each treatment plant has its own "critical concentration" of oxygen, and the degree of its absorption is determined mainly by the nature and concentration of contaminants.

Domestic effluents are a relatively weak nutrient solution. In it, the rate of oxygen absorption exceeds the rate of absorption of nutrients, so oxygen is rarely limited in facilities that treat such wastewater [4].

In concentrated industrial wastewater with a high content of easily oxidizable organic substances, the rate of absorption of nutrients by bacteria will exceed the rate of absorption of oxygen, which, in this case, is limited. Thus, the required degree of aeration should first consider the load on the pollutants rather than hydraulic loads. Increasing the content of dissolved oxygen in aeration tanks to more than 3.5–4.0 mg/ dm^3 has little effect on the efficiency of the biochemical oxidation of pollutants but greatly increases energy costs [5].

Poor aeration conditions for activated sludge can be caused by the following reasons: reduction of the amount of supplied air, destruction and clogging of the elements (filter plates, perforated pipes, fine-bubble dispersants, etc.) that supply air, aeration zone and all cleaning units; increasing the specific loads on the activated sludge due to the increase in the content of dissolved organic substances in the waters entering the treatment; the effect of toxicants on activated sludge (toxicants block respiratory enzymes in the protozoa and multicellular organisms of activated sludge or respiratory pigments in bacteria); increase in oxygen absorption of activated sludge due to violation of the regime of unloading sludge from secondary settling tanks; exceeding the optimal concentration of return sludge (lack of oxygen occurs with increasing biomass of activated sludge) [6].

Aeration conditions can be improved by adjusting the technological mode of operation (limited possibilities) and increasing the percentage of oxygen use of activated sludge by changing the aerating elements. With large-bubble aeration, the size of the openings in the pipes are 5–6 mm, and the use of oxygen by activated sludge is 6–7%, which does not create an ideal mass transfer of dissolved oxygen from the liquid into the bacterial cell. When the size of the holes in the air supply pipes is reduced to 2–2.5 mm, the use of oxygen increases to 8–12%, and when small-bubble diffusers are used (the size of the holes is 200–500 μm) to 15–18%. Aeration systems used in aeration tanks are divided into three types: pneumatic, mechanical and combined. During operation, calculations are necessary to assess the effectiveness of the aeration system used.

Calculations are provided for the most commonly used pneumatic system (mechanical aeration system; calculations are performed on the cost of electricity).

The air consumption for the treatment of 1 m ^{3} wastewater ($D_{1m}{}^3$, m^3/m^3) is calculated by the formula

$$D_{1m^3} = \frac{D_{\text{average daily}}}{Q_{\text{average daily}}}, \tag{5.3}$$

where $D_{average\ daily}$—average daily air consumption, m^3 day; $Q_{average\ daily}$—the average daily wastewater flow, m^3/day.

Example

$$D_{1m^3} = \frac{21{,}294\ \text{m}^3/\text{day}}{2340\ \text{m}^3/\text{day}} = 9.1\ \text{m}^3/\text{m}^3.$$

The air consumption for the treatment of m^3 wastewater can be from 3.5 to 15 m^3/m^3.

The air consumption per 1 kg of removed organic matter, characterized by the indicator BOD_5 (D_{BOD5}, m^3/kg), is calculated by the formula

$$D_{BOD_5} = \frac{D_{1\mathrm{m}^3}}{L_{IN} - L_{OUT}}, \tag{5.4}$$

where L_{IN}—the content of BOD_n in wastewater entering treatment, mg/dm^3 or kg/m^3; L_{OUT}—the BOD_5 in treated wastewater, mg/dm^3 or kg/m^3; $D_{1\mathrm{m}^3}$—the air consumption for treating 1 m^3 of wastewater, m^3/m^3.

Example

$$D_{BOD_5} = \frac{9.1\ \mathrm{m}^3/\mathrm{m}^3}{0.214\frac{\mathrm{kg}}{\mathrm{m}^3} - 0.013\,\mathrm{kg/m}^3} = 45.2\ \mathrm{m}^3/\mathrm{kg}.$$

The air consumption per 1 kg of removed organic matter, which is characterized by the indicator BOD_5, averages from 30 to 55 m^3/kg and can increase up to 90 m^3/kg during wastewater treatment at high-load facilities and reach significant costs up to 120 m^3/kg of removed BOD in the process of prolonged aeration.

The aeration intensity J, m^3/m^2 h, is calculated by the formula

$$J = \frac{D_{1m^3} \cdot H}{T}, \tag{5.5}$$

where J—the depth of aeration tanks or regenerators, sand traps, mineralizers, m; T—aeration period, hours

Example

$$J = \frac{9.1\ \mathrm{m}^3/\mathrm{m}^3 \cdot 4.4\ \mathrm{m}}{7.5\ \mathrm{h}} = 5.3\ \mathrm{m}^3/(\mathrm{m}^2 \cdot \mathrm{h}).$$

The intensity of aeration must be sufficient to ensure both the saturation of the sludge mixture with dissolved oxygen and the mixing process.

In order to calculate the percentage of air oxygen use, it is necessary to consider the total oxygen consumption spent on biochemical oxidation of organic pollutants, nitrification and oxygen saturation of treated wastewater discharged into a natural reservoir.

Similar calculations can be performed, considering the airflow rate for the performance of blowers. Air consumption $D_{1\mathrm{m}^3}$ (m^3/m^3) will be

$$D_{1m^3} = \frac{N}{Q_{\text{average daily}}}, \tag{5.6}$$

where N—the daily capacity of blowers, m³/day; $Q_{average\ daily}$ is the average daily wastewater flow rate, m³/day.

The air consumption per 1 kg of removed organic matter, characterized by the indicator BOD_5 (D_{BOD5}, m³/kg), is calculated by the formula

$$D_{BOD_5} = \frac{N}{L_{util}}, \tag{5.7}$$

where L_{util} ($L_{util} = L_{in}$—L_{out})—the amount of utilized organic matter, which is characterized by an indicator of BOD_5 per 1 day, kg.

Example For 1 day, the airflow was 240,000 m³. Processed organic matter based on BOD_5 is 4000 kg, then

$$240,000\ \text{m}^3 : 4000\text{kg} = 60\ \text{m}^3 \text{per 1kg BOD}_5.$$

The intensity of aeration according to the formula

$$J = \frac{D_1}{F_{aer.fac}}, \tag{5.8}$$

where D_1—hourly air consumption, m³/h; $F_{aer.fac}$—water surface area of aeration facilities, m².

Example

$$F_{aer.fac} = 2572\ \text{m}^2,\ D_1 = 10,008\ \text{m}^3/\text{year},$$
$$J = 10,008\ \text{m}^3/\text{year} : 2572\ \text{m}^2 = 3.89\ \text{m}^3/(\text{m}^2\ \text{year}).$$

The aeration intensity per volume (W, m³) of aeration facilities at W = 12,860 m³ will be

$$10,008\ \text{m}^3/\text{year} : 12,860\ \text{m}^3 = 0.78\ \text{m}^3/(\text{m}^3\ \text{year}).$$

The calculation of the percentage of oxygen use per 1 kg of removed organic matter, characterized by the indicator BOD_5, is given by the following example.

Example Use 60 m³ of air per 1 kg of BOD_5. Convert the volume of air into units of mass

The mass of 1 m³ air at t = 0 °C and a pressure of 760 mmHg will be 1.29 kg/m³, so the mass of 60 m³ air will be

$$60\ \text{m}^3 \cdot 1.29\ \text{kg/m}^3 = 77.4\ \text{kg}.$$

The air contains 20.99% of oxygen; hence, for 1 kg of organic matter removed, which is characterized by a BOD_5, oxygen is used

$$77.4\,\text{kg} \cdot 0.21 = 16.3\,\text{kg}.$$

Calculate the percentage of air oxygen per 1 kg of organic matter removed, which is characterized by an indicator of BOD_5:

$$\frac{1\ \text{kg}}{16.3\ \text{kg}} \cdot 100\% = 6.13\ \%.$$

Daily increase (In, mg/dm^3), the amount of sludge (due to the removal and absorption of pollutants from the water), which is newly formed, for municipal and industrial wastewater with a similar composition is determined by the formula

$$In = Q \cdot 0.8B_{\text{clar}} + 0.3L_{\text{clar}}, \tag{5.9}$$

where Q is the average daily volume of treated wastewater, m^3/day; B_{clar}—the content of suspended solids in clarified water after the primary settling tanks, mg/dm^3; L_{clar}—BOD_5 in clarified waters after the primary settling tanks in the shaken sample, mg/dm^3.

There are several formulas for calculating sludge growth based on biomass balance; they take into account wastewater consumption, load, biomass accumulation rate, sludge self-oxidation rate constant, etc. Most of all, these formulas are suitable only for aeration tank mixers because for aeration tank displacers, it is difficult to calculate due to changes in processes as wastewater progresses from the beginning to the end of the aeration tank. This should be kept in mind; however, the sludge growth at the Kamyshin treatment facilities with aeration tanks-displacers will be calculated using one of the following formulas

$$In = Q \cdot Y(L_{in} - L_{out}) \cdot 10^{-3}\ \text{kg/g}, \tag{5.10}$$

where *In* is the increase in activated sludge, kg/day; Q—wastewater flow, m^3/day; Y—a coefficient accounting for the conversion rate of utilized organic matter into biomass, kg/kg (it can be considered as a dimensionless value in calculations); L_{in}—the organic matter content, characterized by the BOD_5 indicator, in wastewater entering treatment, g/m^3; L_{out}—the same in treated wastewater, g/m^3; 10^{-3} kg/g—a constant coefficient converting grams to kilograms.

The coefficient (Y, g/g), taking into account the rate of conversion of utilized organic matter into biomass, is calculated by the formula

$$Y = \frac{N}{1 + k_d \cdot B}, \tag{5.11}$$

where N—the load on BOD_5 per unit of sludge, kg/kg; B—sludge age, days; k_d is the rate constant of endogenous oxidation, ranging from 0.025 to 0.075 days, usually taken as 0.06 days in calculations.

For aeration tanks, it is necessary to consider the result obtained by formula (5.9) to be more accurate. However, it should be adjusted to a lower value taking into account the self-oxidation of sludge and its removal from secondary settling tanks. From the given examples, it is clear that there are no perfect formulas for calculating sludge growth; at the same time, the quantitative characteristic of the growth is very important during the operation of facilities, as it is used to determine the amount of biomass removed for processing and utilization. In operation, the increase in sludge is calculated by summing the masses of sludge removed from the system, i.e. taken out with purified water and pumped to the sludge compactor or other sludge processing facilities, as well as lost in the process of self-oxidation. The calculated values of sludge growth are used as approximate.

Due to the process of sludge regeneration, if it is adequately maintained, the oxidizing properties of activated sludge are restored, and the adverse effects on sludge are mitigated. This includes the inactivation of: toxicants that are inert to biooxidation but accumulated in the sludge (metals, pesticides, etc.); toxicants that are fully or partially oxidized (SPAR petroleum products); bioavailable but difficult-to-oxidize substances sorbed on sludge (phenols, etc.) [7].

All these substances are either oxidized or deposited and inactivated in the sludge during regeneration. The mechanism of regeneration is not fully understood. However, it is clear that by eliminating the process of sorption of pollutants on the sludge (wastewater is not fed into the regenerator), the supply of oxygen, occurs [8]:

- Restoration of toxic pigments and sludge enzymes blocked by toxicants.
- Deposition in the biopolymer gel of inert toxicants.
- Sequential hydrolysis and consumption of oxidized organic matter due to the extended residence time of sludge in the system.

5.2.3 *Recirculation Sludge. Degree of Recirculation*

The flow rate of wastewater through the construction of biological treatment is such that a significant part of the activated sludge is carried out with the treated water into the reservoir. Removal is calculated considering the volume of treated wastewater and the amount of suspended solids in the treated water. To continuously maintain the required amount of activated sludge biomass in aeration tanks, partial recirculation is required—the return of biomass from the settling tank to the aeration tank, which allows ensure conditions when the dilution rate does not exceed the growth rate of microorganisms [9].

The volume of return sludge removed from the secondary settling tanks and sent to the regenerator is 30–70% of the volume of treated wastewater. For each treatment plant, this indicator is individual and is determined by calculation as the recirculation ratio:

$$R = \frac{\alpha_{\text{average}}}{\frac{1000}{I} - \alpha_{\text{average}}} \cdot 100\,\%, \tag{5.12}$$

where R is the recirculation ratio of the return sludge to the flow rate of treated wastewater; I—sludge index (determined in a sample taken at the end of the aeration zone), cm^3/g

In formula (5.12), the coefficient 1000 converts cubic decimeters to cubic centimeters and has a dimension of cm^3/dm^3; α_{average}—the average dose of sludge, g/dm^3, calculated considering the dose by sludge mass in all corridors of aeration tanks and regenerators:

$$\alpha_{\text{average}} = \frac{W_1 \cdot \alpha_1 + W_2 \cdot \alpha_2 + W_3 \cdot \alpha_3}{W_1 + W_2 + W_3}, \tag{5.13}$$

where W_1, W_2, W_3—volumes of corridors of aeration tanks and regenerators, m^3; α_1, α_2, α_3—dose of activated sludge in each corridor, g/dm^3.

Example

$$R = \frac{2.0\ \text{g/dm}^3}{\frac{1000\ \text{cm}^3/\text{dm}^3}{100\ \text{cm}^3/\text{g}} - 2.0\ \text{g/dm}^3} \cdot 100\ \% = 25\ \%.$$

The value of R is 25% of the total volume of treated wastewater.

The given example shows that the lower the average dose of sludge and the higher the sludge index, the larger the volume of sludge that must be returned to the regenerators. Thus, if the specific loads on the sludge increase and its sedimentation properties deteriorate until the sludge bulking, the removal of sludge from the secondary settling tanks increases, ie the sludge losses increase, so it is necessary to return the maximum amount of sludge to the system. If the facilities maintain a lower percentage of recycling than estimated, the sludge in the secondary settling tanks is compacted and will rot, and in aeration tanks, specific loads will increase due to the low concentration of sludge.

If a higher recirculation rate is maintained, the energy consumption for sludge pumping will increase, and the sludge in the secondary settling tanks will be under-compacted. In both cases, the removal of suspended solids from the secondary settling tanks will increase, and the quality of cleaning will deteriorate.

To calculate the age of sludge (AS), several formulas are proposed (it is expressed in any case by day), the most used of which determines the age as a fraction of the total amount of sludge and its growth per day:

$$AS = \frac{\alpha_{average} \cdot \text{T}}{In \cdot 24} \cdot 1000, \tag{5.14}$$

where In—the increase in sludge, g/dm^3; T—aeration period, h; and α_{average}—the average dose of sludge, g/dm^3; the coefficient of 1000 translates grams to milligrams and has a dimension of mg/g.

If the sludge increase is given in g/dm^3, then in formula (5.14), the coefficient 1000 is absent.

Example Calculation of the age of sludge at the wastewater treatment plant with the treated wastewater flow rate 95,170 m^3/day, the average dose of activated sludge 1.1 g/dm^3, the aeration period of 8.9 h, the average daily gain of sludge 114.04 mg/ dm^3, or 0.114 g/dm^3.

$$AS = \frac{1.1\ \mathrm{g/dm^3} \cdot 8.95\ \mathrm{h}}{0.114\ \mathrm{g/dm^3} \cdot 24\ \mathrm{h/day}} = 3.59\ \text{days}. \tag{5.15}$$

This means that 27.8% (100%: 3.59 days) of activated sludge from the total mass is renewed at the treatment plant in 1 day. Such a renewal percentage is quite high and is achieved due to the low sludge age. It is necessary to update no more than 12–13% of the total mass of sludge on the facilities operating in the mode of complete oxidation of pollutants and nitrification.

The formula for calculating the age of the sludge (AS), taking into account the consumption of excess sludge, takes the following form:

$$AS = \frac{\alpha_{average} \cdot W}{(Q - q_{\mathrm{excess}}) \cdot B + q_{\mathrm{excess}} \cdot \alpha_{\mathrm{excess}}}, \tag{5.16}$$

where W—the volume of aeration structures, m^3; Q—wastewater flow, m^3/day; q_{excess}—consumption of excess sludge, m^3/day; B—the content of suspended solids in purified water after secondary settling tanks, kg/m^3 or mg/dm^3; $\alpha_{average}$—the average dose of sludge, g/dm^3 or kg/m^3; α_{excess}—concentration of excess sludge, mg/dm^3 or kg/m^3.

From formula (5.16), it follows that the age of the sludge can be controlled by changing the consumption of excess sludge.

Example The consumption of excess sludge at the wastewater treatment plant is 3200 m^3/day, the content of suspended solids in the treated water is 10 mg/dm^3 (10^{-2} kg/m^3), the concentration of excess sludge—3 g/dm^3 (3 kg/m^3):

$$\begin{aligned} AS &= \frac{1.1\ \mathrm{g/dm^3} \cdot 35,500\ \mathrm{m^3}}{\left(95,170\ \mathrm{m^3/day} - 3200\ \mathrm{m^3/day} \cdot 10^{-2}\ \mathrm{kg/m^3} + 3200\ \mathrm{m^3/day} \cdot 3\ \mathrm{kg/m^3}\right)}. \\ &= 3.7\ \text{days} \end{aligned}$$

From the above calculations, it can be concluded that the dose of activated sludge is kept low, and the volume of sludge shipped from the system is higher than optimal. As a result, the sludge age is greatly underestimated for these facilities.

In operating biological treatment facilities, the optimal age of activated sludge is determined not only by calculation but also by the results of observations on the efficiency of wastewater treatment in terms of maintaining different ages.

5.3 Anaerobic Waste Recycling Systems

5.3.1 *Overview of the Biogas Technology Industry in Global North and Global South*

The increasing concentration of carbon dioxide in the atmosphere, a primary driver of the greenhouse effect, stems largely from the prolonged reliance on traditional non-renewable energy sources. These resources are being depleted at an accelerating rate, while their extraction is becoming increasingly complex and economically less viable due to rising costs [9].

The discovery of gas hydrates or methane hydrates in submarine permafrost zones—such as those in the Caspian, Black, Mediterranean, Okhotsk, and Japanese seas—has raised the prospect of temporarily mitigating global energy shortages. However, the large-scale exploitation of these resources would likely exacerbate climate change by releasing additional greenhouse gases into the atmosphere.

To address the challenges of resource depletion, environmental pollution, and the growing global population, the focus of energy production has shifted toward sustainable and energy-efficient technologies. Renewable energy sources, particularly biomass energy, have gained significant attention. Biomass can be derived from livestock and agricultural waste, wood residues, plant biomass, marine phytoplankton plantations, and photobioreactor-generated materials.

In alignment with the Kyoto Protocol [10], several countries—including Brazil, the UK, Denmark, Germany, Ireland, Sweden, Finland, and the USA—have prioritized research on bio-energy plantations. These initiatives aim to capture atmospheric carbon dioxide while simultaneously producing biomass for energy generation. According to scientific estimates, the global technical potential of biomass energy could reach 500 exajoules per year by 2050, roughly equivalent to the total biomass-derived heat energy currently produced in agriculture and forestry.

The expansion of energy crop plantations presents a promising avenue for biofuel production. Ukraine, for example, has the potential to allocate 27 million hectares for high-yield energy crops, while the USA could expand such plantations to 77 million hectares. Brazil, the world's largest producer of bioethanol, currently cultivates over 7 million hectares of sugarcane for energy purposes and could dedicate an additional 58 to 148 million hectares without compromising existing forests. Similarly, Indonesia has the capacity to allocate 27 million hectares of oil palm for bioenergy production.

A key challenge in biofuel production is achieving an optimal balance between carbon dioxide reduction and energy generation. The energy output from biofuels must exceed the energy required for their production to ensure a positive energy balance. While some biofuel production processes yield net-zero or even negative energy balances, they can still provide environmental benefits when integrated with clean energy sources at the point of consumption. Many countries with substantial reserves of traditional energy resources—such as the USA, Canada, Germany, Brazil, and India—have already made significant technical advancements in biofuel production and continue to expand the bioenergy sector.

The benefits of bioenergy are well recognized and include:

Reduction of greenhouse gas emissions—mitigating climate change by lowering carbon footprints.
Sustainable development—utilizing clean and renewable energy sources.
Decreased dependence on fossil fuels—enhancing energy security.
Versatile applications—bioenergy can be used for electricity, heating, transportation, and industrial processes.
Agricultural and rural economic growth—creating new opportunities for farmers and industries.
Lower production costs and improved competitiveness—increasing market viability.
Enhanced national security—particularly for countries with limited fossil fuel resources.

Anaerobic digestion, a process in which organic biomass is converted into biogas (a mixture of hydrogen and biomethane) and organic fertilizers (such as biocomposites or digestate), is gaining importance as a sustainable energy solution [11]. The composition and yield of biogas depend on the feedstock and technological processes used. The main raw materials are energy crops grown specifically for this purpose (corn, cereals, sugar beet, and many others) and agricultural and livestock waste. These include organic wastes from food processing (e.g. press cake, grease trap waste), vegetable waste from wholesale markets, food waste, cut grass, material from landscape conservation activities, and organic waste from household landfills. The utilization of biogenic waste for biogas production solves the problem of its accumulation in areas and the contamination of the air and soil environment with substances generated by the decomposition of organics under inadequate disposal.

Different substrates yield varying levels of biogas productivity and methane content. More than 60 biogas technologies are currently in use or under development worldwide. Comparative analyses show that producing biogas from whole-plant biomass is more energy-efficient than biodiesel (which relies only on seeds) or bioethanol (which utilizes carbohydrates). Additionally, the byproducts of anaerobic digestion—liquid and solid residues—serve as nutrient-rich organic fertilizers. Calculations show that using all the energy plant biomass for producing biogas, compared to the production of biodiesel (uses only seeds) and bioethanol (uses carbohydrates), is highly energy efficient. Furthermore, liquid or solid residues from digestion can be used as organic fertilizer due to their high nutrient content.

The energy production from biogas is unrelated to the time of day, season, or weather conditions. Therefore, it can be produced continuously depending on the demand. In Germany, due to the fixed tariff for electricity from biogas, electricity and heat are produced near the biogas plant, which is the main way of using biogas. In the case of long distances between the place of production and use, biogas can be transported via gas pipelines.

Biomethane can be separated from other associated gases from biogas by wet purification under pressure, variable pressure adsorption, physical and chemical purification processes, and membrane technology.

Biomethane production appeared to be the focus of researchers and developers in many countries. This is because it significantly increases the energy efficiency of this type of gas production process by using whole plant biomass to produce biomethane and the opportunity to supply it to a common gas network with different uses.

Incomplete combustion of biomass in gas-producing furnaces produces generator gas: a mixture of flammable carbon monoxide (26–30%) with hydrogen (13–15%) and non-flammable nitrogen (45–55%) and carbon dioxide (5–8%). The heating value of generator gas varies between 10–13.4 MJ/m^3. Generating gas can be used to produce heat, electricity, and liquid fuels as it does not require sophisticated equipment.

Gasification of coal with steam conversion and partial methane oxidation produces a synthesis gas—a mixture of carbon monoxide and hydrogen—with a ratio of $CO:H_2$ gases from 1:1 to 1:3, depending on the production method.

Biogas production is an efficient and investment-attractive technology due to the significant raw material resource potential, favourable natural and climatic conditions, and low-cost price of this type of energy.

Biogas plants can be set up as waste treatment facilities on farms, alcohol factories, sugar refineries, and meat processing plants. The biogas plant can replace a veterinary sanitary plant, which means that carrion can be utilized in biogas instead of producing meat and bone meal.

Among industrialized countries, Denmark has the leading position in the production and use of biogas according to relative indicators (biogas represents up to 18% of the total energy balance). Based on absolute indicators for the number of medium and large plants, Germany leads with 8000 plants. In Western Europe, at least half of all poultry farms are heated with biogas.

The bioenergy research topic is relevant in agrarian countries and countries with high amounts of organic waste. The United States of America takes the lead in the research on this topic and the number of publications—more than 100 scientific papers; then there comes China—60 papers; Germany and England—more than 50 papers; and Sweden—more than 20 papers (Fig. 5.4).

The largest number of scientific studies are focused on improving the environmental situation in the regions and the energy sector, biofuel production, and chemical engineering by producing various useful chemical additives (Fig. 5.5).

5.3.1.1 Modeling the interconnections between bioenergy trend clusters as an integrated solution for environmental protection.

4 clusters were formed based on Scopus database data using VOSviewer cluster software (Fig. 5.6):

(1) Cluster (red) reveals the environmental problems of changing the direction of implementation of stationary energy sources with the development of bioenergy potential and the creation of strategies for this development at the level of regions.

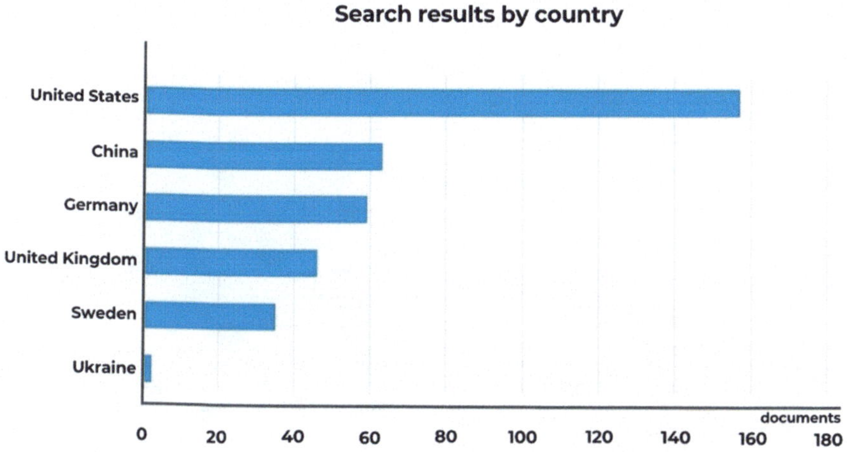

Fig. 5.4 Analysis of search results by country [12]

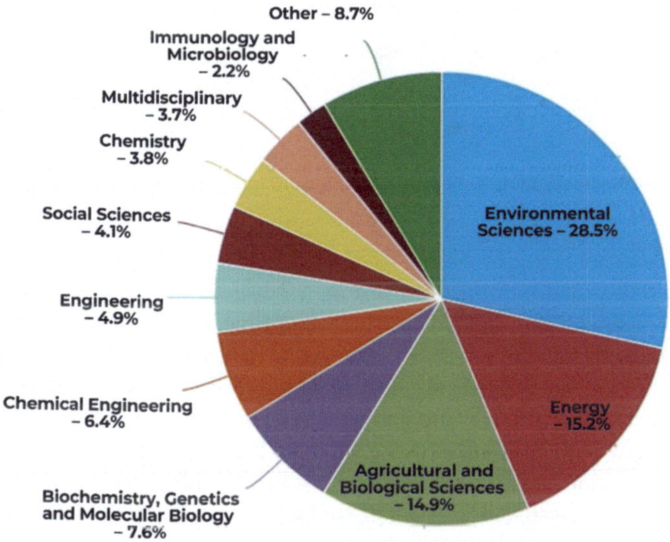

Fig. 5.5 Analysis of search results by topic of research in the bioenergy technology field [12]

(2) Cluster (yellow) covers the restoration of ecological systems, particularly forests and reducing CO_2 emissions from bioenergy.
(3) Cluster (green) covers the production and use of different types of fuel and energy produced by introducing and improving bioenergy technologies.
(4) Cluster (blue) covers the impact of bioenergy technologies on environmental restoration and purification and reduction of damage from anthropogenic impact.

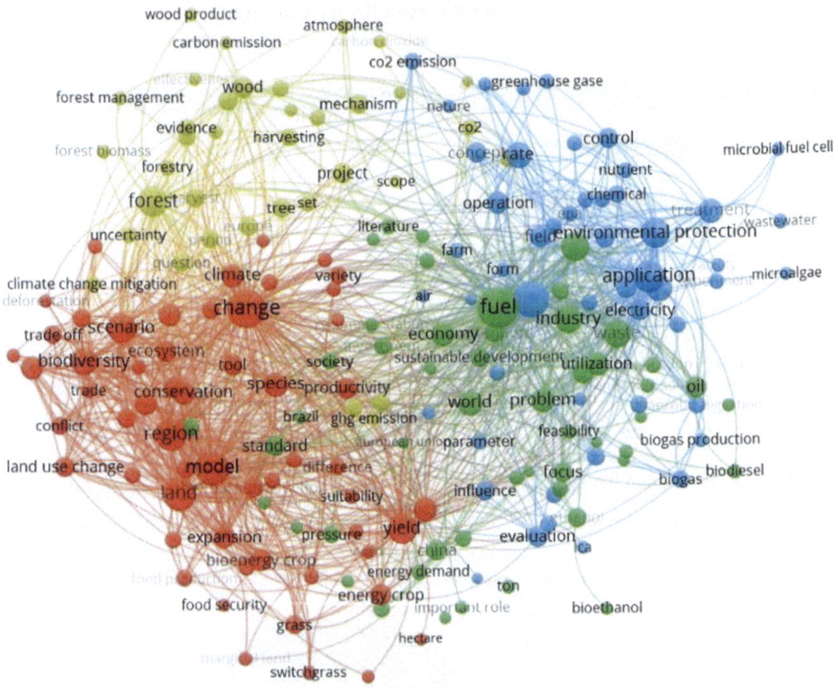

Fig. 5.6 Network visualization: 4 clusters, 11,545 links, 27,577 total link strength (Scopus database data) [12]

The overlay visualization shown in Fig. 5.7 was identified as a more effective tool to verify the latest trends in research on a time scale. The size of the circles corresponds to the predominance of the period when the study in this area was published. The colour depends on the year of publication (average for the cluster), with the latest year's colour being yellow. Thus, the terms "fuel" and "change" are used most intensively in connection with the publication year.

The visualization figure shows that biogas is being studied in various applications. There are various classification systems; for example, [13] designed a typical biogas system with 3 utilization pathways: biogas combined heat and power, biogas solid oxide fuel cells, and biogas upgrading. Based on the system assessment results, the biogas upgrading pathway has the highest systematic energy efficiency and the shortest payback period. Current biogas upgrading techniques are systematically discussed in Khan et al. [14]. The review showed a need for further research on methane (CH_4) loss, environmental effects, energy consumption, and economic assessment challenges.

One of the study's main clusters is changing the direction towards bioenergy and creating new development models. Bioenergy is well developed in EU countries. According to the analysis by Banja et al. [15], there are movements towards a capacity market mechanism to support biogas in the electricity sector. This slows the pace

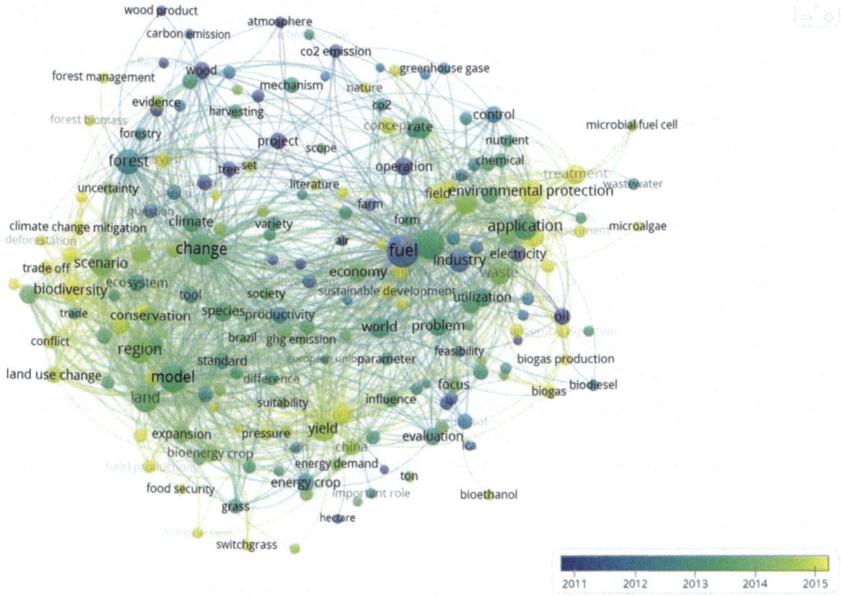

Fig. 5.7 Overall visualization showing the unit of measurement as the time since publication (Scopus database data) [12]

of biogas introduction in the EU biomass electricity markets. Several EU countries continue to keep the dynamics of their biogas electricity markets alive. Still, biogas faces stiff competition with solar and wind energy sources and inexpensive natural gas, so further modernization of the biogas and biomethane production process is needed. According to Achinas et al. [16], investments in anaerobic digestion are expected to succeed due to the low cost of available feedstocks—lignocellulosic sources (manure, fruits, vegetable waste) and the wide range of biogas applications. Research initiatives based on the article focused on improving anaerobic digestion control and its efficiency, i.e., microbial activity is a crucial parameter for process stability and biogas yield, and further research is required. The article by Ferdeş et al. [17] overviewed the enzymatic and biological pretreatment of feedstock for biogas production. Biological pretreatment is a highly effective method for enhancing biogas production, and the least expensive treatment is applied to lignocellulosic biomass. In the future, increasing the efficiency of enzyme preparations and the capacity of microorganisms to transform the substrate could be achieved by selecting new highly productive microbial strains and using molecular genetic techniques.

The impact of bioenergy technology solutions is critically discussed in Paolini et al. [18]. This paper shows that biogas can significantly contribute to the abating greenhouse gas emissions but also provides the conditions for undesirable methane and N_2O emissions (they are scarcely related to direct release from biogas combustion, whilst biomass storage and digestate management are the critical steps handling). According to the article, digesters are preferable to untreated biomass in the medium-short term. Upgrading to biomethane can generally improve air quality and reduce greenhouse gas emissions. However, methane losses in the off-gas may affect the sustainability of the whole process.

Based on an analysis of current research trends, the scientific focus is on improving biofuel production and the environmental situation in the regions and energy sector. After comparing international studies with local research in the same field of methane fermentation intensification, we can see that international studies are more complex and cover the overall system.

The analysis of patent databases of technological solutions for biogas production highlighted the following directions of development: facilities for biogas production, bioreactors for different fermentation phases, machines for biological desulphurization of biogas, methods, and systems for solid waste processing. Based on cluster analysis, the most common research topics are changes in bioenergy and its development models, reforestation, biofuel production and application, and the environmental impact of bioenergy technological solutions.

5.3.2 Trends in the Development of Biogas Plants

Along with the improvement of aerobic biological treatment methods, the technology of anaerobic treatment, used mainly for highly concentrated wastewater, is developing intensively. The rapid development of anaerobic technology is associated with the desire to create compact and efficient devices that are reliable and flexible, with low capital operating and energy costs for wastewater treatment.

Installations for biomethanogenesis, considering their volume and productivity, can be divided into several categories: reactors for small farms in rural areas (1–20 m^3); reactors for farms in developed countries (50–500 m^3); reactors for processing industrial effluents (alcohol, sugar industry) (500–10,000 m^3) and reactors for processing solid waste from urban landfills ($1–20 \times 10^6$ m^3). Metal or reinforced concrete methane tanks can have different shapes, including cubic and cylindrical. The designs and details of these plants vary somewhat, mainly due to the type of raw materials processed. There is a wide variety of installations for the implementation of the process of methanogenesis, the structural details and layout of which are determined by the priority of the task to be solved in a particular process: waste disposal and wastewater treatment or obtaining biogas of the required quality. Thus, among the plants operating in developed countries, there are both medium and large units (digests) equipped with devices for purification and compression of biogas, electric generators and water purifiers. Such plants can be part of complexes with industrial enterprises (sugar processing, alcohol or dairy enterprises), sewage stations or large specialized farms. If the primary goal of the process is waste utilization, the system must include a unit for fractionation and separation of large solid particles. Methane tanks can operate in full mixing, full displacement mode as anaerobic biofilters or fluidized bed reactors, and in contact mode. A simple design of a methane tank is an ordinary fermentation pit in the soil with a fixed gas volume.

The most common classification of anaerobic bioreactors is based on the form of macrostructures of methanogenic biomass in them, according to which biogas plants are divided into reactors with suspended sedimentation and attached biomass.

Second-generation anaerobic reactors have become the main facilities for treating highly concentrated wastewater, providing effective wastewater treatment in a very wide range of concentrations (BOD = 0.3–100 g/l), with a water retention time in the apparatus from 30 min up to 48 h. The anaerobic method is used to treat a wide variety of wastewater, primarily for the pulp and paper industry, sugar production and breweries.

Treatment of highly concentrated wastewater in anaerobic conditions compared to aerobic treatment can reduce capital costs by almost 10 times and place facilities in a much smaller area (approximately 10–20 times). At the same time, operating costs compared to aerobic cleaning are reduced almost three times. The disadvantages of the anaerobic method due to the low growth rate of anaerobic bacteria, their high sensitivity to changes in pH, temperature and fluctuations in pollutant concentrations in wastewater, as well as the lower speed of anaerobic processes compared to aerobic are successfully overcome by the indisputable advantages of modern anaerobic systems. Significant advantages of the anaerobic method are the reduction of the amount of excess sludge (3–10 times), its stability, and the possibility of obtaining additional energy due to the biogas produced.

Anaerobic treatment technology can be implemented in reactors with suspended sedimentation and attached biomass. Figure 5.8 shows a diagram of the main directions of development of anaerobic technology; black dotted lines indicate the types of devices that have been developing recently.

The most common classification of anaerobic reactors is based on the form of macrostructures of methanogenic biomass in them. According to this principle, all structures can be divided into reactors with suspended sedimentation (sludge) and attached (biofilm) biomass. Examples of the first type of reactors are traditional methane tanks (anaerobic digesters), anaerobic lagoons, anaerobic contact reactor (ACR), UASB reactors—up-flow anaerobic sludge blanket, EGSB reactors—expanded granular sludge beds and baffled reactors.

The second type includes an anaerobic downflow stationary fixed film reactor (DSFF-reactor), an anaerobic fluidized bed reactor (AFB), and a rotating biological contactor.

Some designs—an anaerobic filter (AF) with upward flow and a hybrid reactor (AF UASB)—combine elements of both types of reactors.

Septic tanks are horizontal settling tanks of closed type, in which the sediment of solid particles form at the bottom rots and decomposes by anaerobic microorganisms (Fig. 5.9).

In septic tanks, activated sludge from secondary settling tanks, sludge from primary settling tanks, and foam are often fermented to reduce their volume and eliminate foul odours and pathogenic microflora. The sediment is kept for up to 4–6 months. Due to the decomposition of organic matter and compaction, its volume is reduced to 50%.

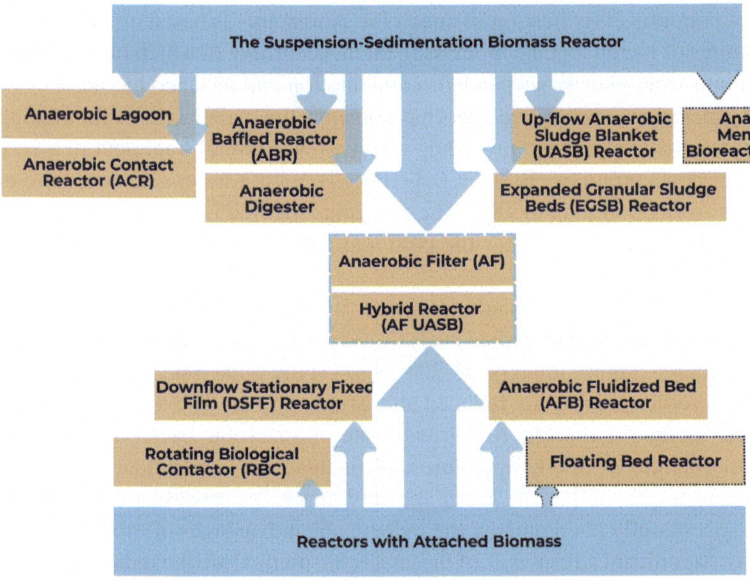

Fig. 5.8 Classification of modern anaerobic technology (black dotted lines—the types of devices that have been developing recently; blue dashed line—applies to both categories)

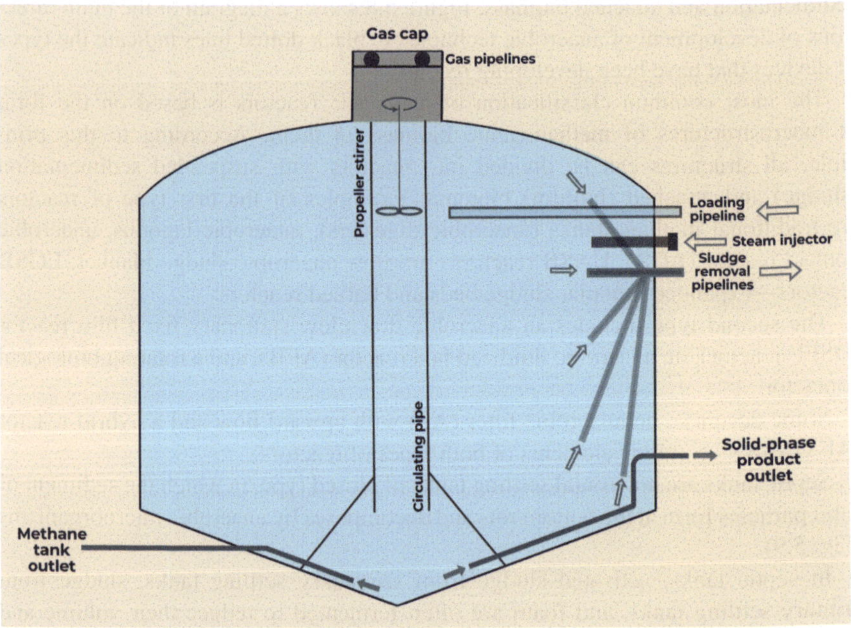

Fig. 5.9 Schematic diagram of a septic tank system

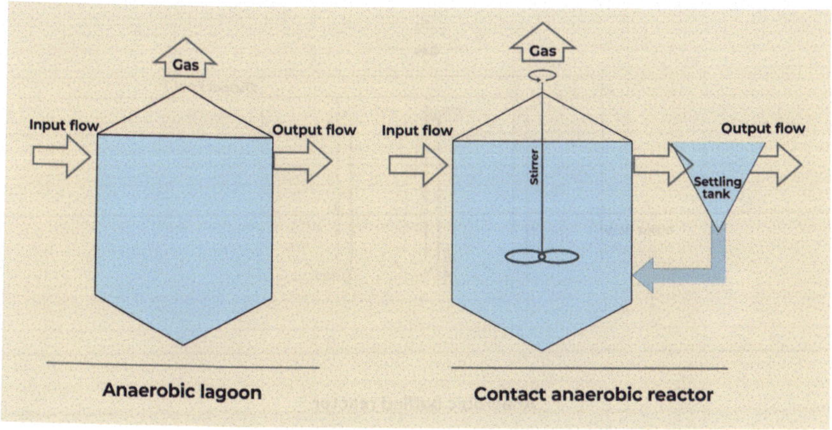

Fig. 5.10 Schematic diagram of anaerobic lagoon and anaerobic contact reactor (ACR)

In **methane tanks (anaerobic digesters)**, in contrast to septic tanks, mixing, heating, and control of basic parameters (temperature, composition of raw materials, intensity of loading of the device, etc.) are carried out. The process is more intense than in septic tanks. The released biogas is collected and disposed of. Methane tanks usually operate in a periodic mode of loading waste or wastewater with a constant collection of biogas and unloading solid sludge as the process is completed.

In **anaerobic lagoons**, waste is stabilized using natural processes (Fig. 5.10).

Anaerobic contact reactor is comparable to aeration tanks with an aerobic sludge activation system (Fig. 5.10). Used for the treatment of relatively diluted domestic wastewater with a COD value of approximately 1300 mg/l. An anaerobic lagoon has a laminar flow without mixing, while mixing occurs in a contact anaerobic reactor. In addition, the contact anaerobic reactor provides for the return of part of the activated sludge after anaerobic digestion as a starter to intensify the process.

The anaerobic baffled reactor (ABR) (Fig. 5.11) is a rectangular container with a series of compartments divided by parallel vertical baffles. The flow moves sequentially from bottom to top and then from top to bottom, passing in each compartment through a layer of granules or flocs of biomass formed there.

The fluidized bed reactor (Fig. 5.12) contains fine media, such as sand or granular activated carbon, to which bacteria are attached. The relatively high rate of ascent of the wastewater causes the carrier particles to rise to a point where the negative buoyancy of the carrier is equal to the upward force of friction against the water. Then, the fluidized bed reactor is stabilized. Usually, part of the treated stream is returned to the inflow to maintain a high upward velocity, even if the flow rate of wastewater is low [19].

AF is an **anaerobic filter**. UAF is an **upflow anaerobic filter** developed in the late 60 s of the XX century. This system resembles an irrigated filter, where a stone carrier was initially used to attach the biofilm. The anaerobic filter was used to

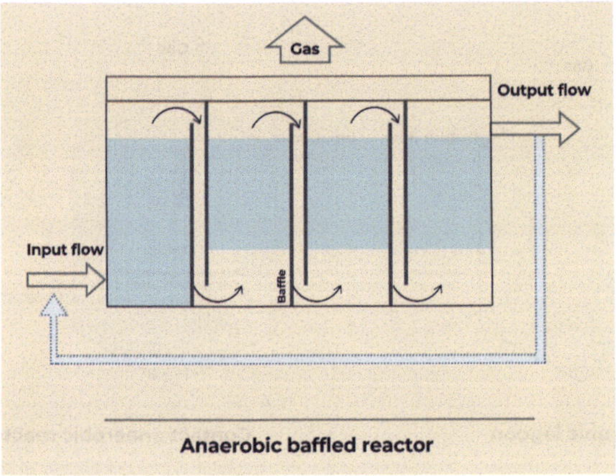

Fig. 5.11 Schematic diagram of anaerobic baffled reactor

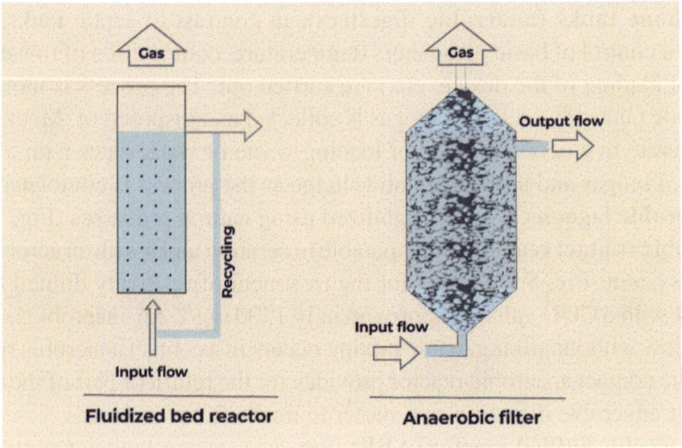

Fig. 5.12 Schematic diagram of fluidized bed reactor and anaerobic filter

clean substrates with COD from 375 to 12,000 mg/l; its retention time was 4–36 h (Fig. 5.12).

The development of anaerobic technology has gone from primitive mixers and anaerobic contact reactors to anaerobic filters and modern **UASB and EGSB reactors**. In the 70's of the twentieth century, studying the processes taking place in second-generation anaerobic reactors with upward flow, Letting found that microorganisms that are part of the methane group can form aggregates—dense granules of size 1–3 mm, which easily settles (Fig. 5.13). These granules mainly consist of methanogens, primarily of the genus Methanosaete, which form dense brush- and ball-like structures. Letting proposed a new reactor design using the open effect with an upward flow of wastewater through a layer of anaerobic sludge [20].

Fig. 5.13 Schematic diagram of upflow anaerobic sludge blanket (UASB) reactor

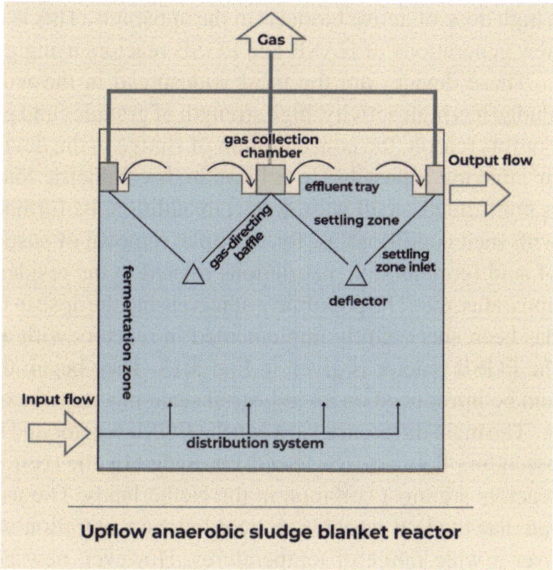

In this design, sludge granulation and its retention are facilitated by using a special built-in gas-sludge separation device located in the upper part of the reactor.

The phenomenon of granulation occurs spontaneously in anaerobic reactors with an upward flow of wastewater containing organic substances that decompose easily (VFAs, sugars) in the absence of toxic components and maintaining optimal living conditions for methanogenic bacteria. In the presence of more protein contaminants in wastewater, inoculation of the reactor with granular sludge is required, intensifying the formation of granules. Above the lower layer is the area of intense turbulent motion of the three-phase system: liquid, sludge and gas. The concentration of sludge in it is 3–10 g/dm^3. There is an intense exchange of sludge particles between the upper and lower layers. The transition of sludge particles from the lower layer to the upper is mainly due to the gas lift effect caused by upward flows of biogas, as well as a result of sludge flotation. The ascending flow is effectively separated in a gas separation device consisting of conical caps of gas collectors and guide baffles-deflectors separating the fermentation zone from the lighting zone. The speed of water movement in the UASB reactor is 0.5–2 m/h. Due to the movement of the liquid and the formed gas bubbles, mixing is provided inside the sludge layer, facilitating the process.

On the phase separation surface under the gas collection cap, the separation of biogas bubbles from the ascending flow of the sludge mixture takes place. Degassing sludge is mainly returned to the fermentation zone. Sewage passes in the gap between the gas caps and deflectors and enters the built-in settling tanks formed by the outer surfaces of the gas caps.

The main direction in the development and improvement of anaerobic reactors, as well as in the case of aerobic bioreactors, is the search for structures that maintain a high dose of active biomass in the apparatus. This is most successfully achieved in new generations of UASB and EGSB reactors using granular activated sludge [21].

These devices are the most widespread in the world in recent years. Granular sludge has high activity, high strength of granules and good sedimentation properties. For this reason, the concentration of sludge in the device's core can reach 50–80 kg/m^3, making it possible to achieve high volumetric loads. However, granular sludge is not formed on all wastewater; in addition, its formation and growth must comply with such conditions as, for example, removal of suspended solids, a certain degree of acid fermentation, restrictions related to the residence time of wastewater in the apparatus, etc. The possibility of increasing the dose in reactors with attached biomass has been successfully implemented in reactors with a fluidized bed. The design of the EGSB reactor is given in Fig. 5.14. They began to actively conquer the market and be introduced on an industrial scale in a number of European countries.

The main difference between the EGSB reactor and the UASB reactor is the higher ascending flow rate (5–12 m/h) provided by the recirculation of the effluent. EGSB reactors are most common in the Netherlands. The use of such a high upward flow rate has made it possible to treat low-concentration wastes from various industries over a wide range of temperatures. However, new generations of UASB reactors

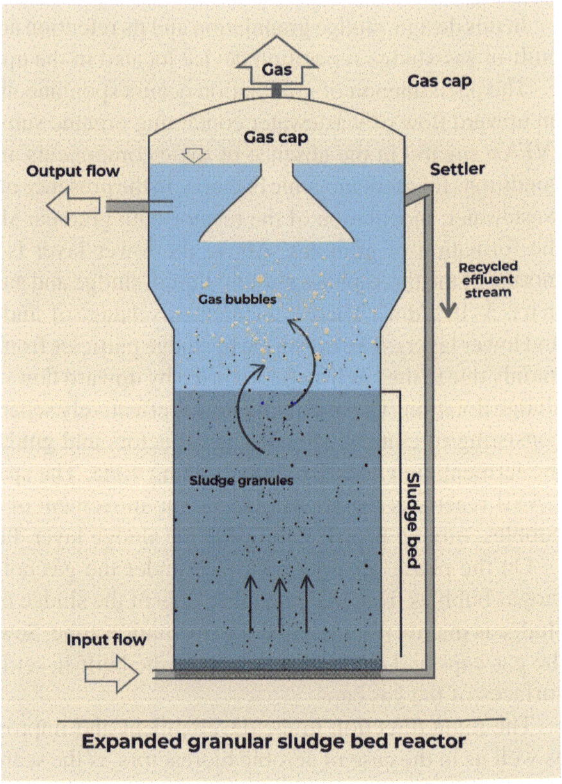

Fig. 5.14 Schematic diagram of the expanded granular sludge bed (EGSB) reactor

are not inferior in their technological parameters and EGSB technology efficiency. UASB technology has the following advantages [22, 23]:

- A regulated hydrodynamic regime and bacterial composition of activated sludge and enzymes provide high productivity for UASB reactors.
- The specific capacity of anaerobic digestion in the reactor (without fermentation accelerators) is up to 16 kg COD/m^3 per day, with a retention time for the fermentation of 6–8 h of dissolved organic pollutants.
- This technological process allows the removal of about 90% of the mass of organic pollution with their concentration by COD at the bioreactor outlet in the range of 1.5–3.0 g/dm^3.
- The reactor can be divided into three parts: sludge base, sludge layer, and three-phase separator.
- The invention of a UASB reactor with a stream flowing through a layer of activated sludge significantly improved the anaerobic methanization of
- the liquid fraction.
- Low biomass growth with anaerobic methanization of the liquid fraction is explained by the methane biocenosis's low energy yield of reactions. Only 8% of energy is spent on biomass growth, 3% is heat loss, and 89% is converted to methane.

The other case design features of the UASB reactor are presented in Fig. 5.15. This scheme represents a development by Shandong Jinhaosanyang, Ltd., a high-tech company specializing in environmental protection equipment. Their UASB reactor technology is widely implemented in various countries, including New Zealand, Finland, the Netherlands, and Australia.

Table 5.2 presents the most important parameters of the main designs of anaerobic apparatus for the mesophilic mode of fermentation and their comparison with the traditional methane tank (anaerobic digester). Table 5.2 shows that an apparatus with a fluidized bed works as efficiently as those with granular sludge.

Biofilter (AF) with upward flow. The biofilter is the first anaerobic reactor with attached biomass. The biomass in the biofilter is retained not only in the form of flocs and granules contained in the cavities of the loading material but also in the form of a biofilm attached to the surface of the carrier. Since the flow of liquid and biogas flows is co-directed, significant mixing in the reactor does not occur, and the hydraulic mode is close to the mode of ideal displacement. In modern installations, mainly planar plastic products are used for loading (previously used bulky materials such as gravel, rubble, slag, etc.) [24].

Hybrid reactors. Numerous studies conducted in the 80 s have convincingly shown that for effective retention of biomass in the AF reactor, it is not necessary to fill its entire volume with loading material: a layer thickness of 25–40% of the working height of the reactor located in its upper part. This not only saves 60–75% of loading material but also eliminates all the disadvantages of classical anaerobic biofilters (siltation of the lower layers, leakage of contaminants). In the reactor, biomass forms both in the lower part where the granular sludge corresponds to the biomass of UASB reactors, as well as on the surface and voids of the carrier. These

Fig. 5.15 Schematic diagram of the UASB reactor developed by Shandong Jinhaosanyang, Ltd

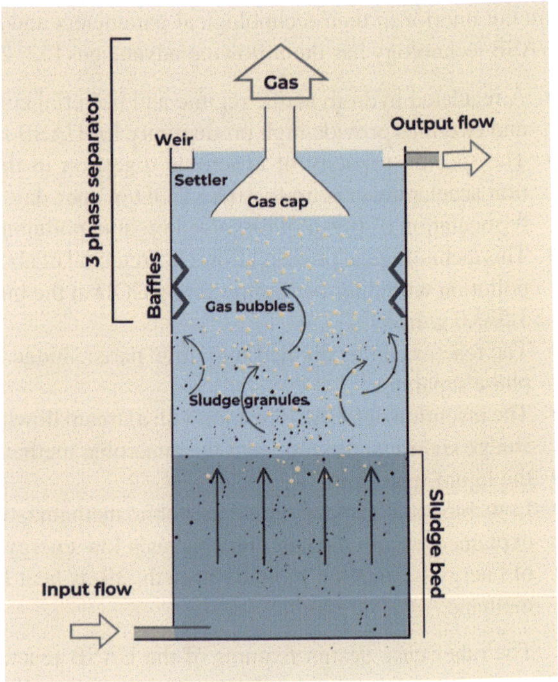

processes increase the efficiency of flocculated sludge retention. Activated carbon and other floating media can be used as the loading material [25].

The internal circulation (IC) reactor can process more waste at lower capital and operating costs. Due to this, it is especially effective in treating highly concentrated wastewater of various chemical compositions with a COD content of 1000 mg/l, for example wastewater from the pulp and paper industry. More than 130 biogas plants have been launched worldwide over the past two decades for wastewater treatment. Preferably, these are combined schemes using UASB reactors in the first stage and aerobic purification in the final stage. So-called three-phase separators are used to retain biomass in the IC reactor (Fig. 5.16). These separators have a small size and a high load of organic matter. At COD 10,000–15,000 mg/dm^3 at high concentrations of organic wastewater, the second-generation UASB reactor has a total loading volume of 5–8 kg COD/m^3; anaerobic IC-reactor of the third generation—volumetric loading rate of 15–30 kg COD/m^3.

The IC reactor consists of two UASB reactors, placed one above the other. The bottom withstands high COD load due to the high concentration of active methane granules; the upper is less loaded. The upper part serves mainly to retain the active granules inside the reactor and the final wastewater treatment and biogas department.

The feature of anaerobic reactors of this type is gas release in two stages. The biogas released in the first conditional UASB reactor creates the effect of a gas lift and thus causes internal circulation of the substrate. Residual separation of biogas

Table 5.2 The most important parameters of the main structures of anaerobic devices

Reactor	Sludge concentration in the apparatus, g/dm^3	Specific surface loading, m^2/m^3	Lower limit of COD in effluent, g/dm^3	Volumetric loading, kg COD/m^2 day	Retention time, h
Traditional anaerobic digester	0.5–3	–	10	0.5–5	192–240
Anaerobic contact reactor (ACR)	5–10	–	2–3	3–8	24
UASB (upflow anaerobic sludge blanket)	20–40	–	0.3	10–25	2–3
EGSB (expanded granular sludge bed)	25–40	–	0.3	30–40	1–2
Anaerobic biofilter	5–20	70–300	0.3	10–15	8–12
DSFF (downflow stationary fixed film)	3–15	60–200	1–2	10–12	24
Hybrid	20–30	70–300	0.3	15–25	2–3
Anaerobic fluidized bed (AFB) reactor	10–40	1000–3000	0.3	30–40	0.5

from the purified liquid occurs in two phases in the separator. The organic load level for an IC reactor is usually twice as high as that of a UASB reactor.

Microbiological studies of methane granules selected during the period of unsatisfactory operation of the installation showed that the content of methanogens in them is less than 1%, and the bulk are acidogenic microorganisms. Obviously, due to unfavourable operating conditions, the symbiotic balance was disturbed: unpretentious acidogenic organisms survived, and methanogens, sensitive to fluctuations in the composition and acidity of wastewater, temperature, volatile fatty acid content, etc., almost entirely died [26, 27].

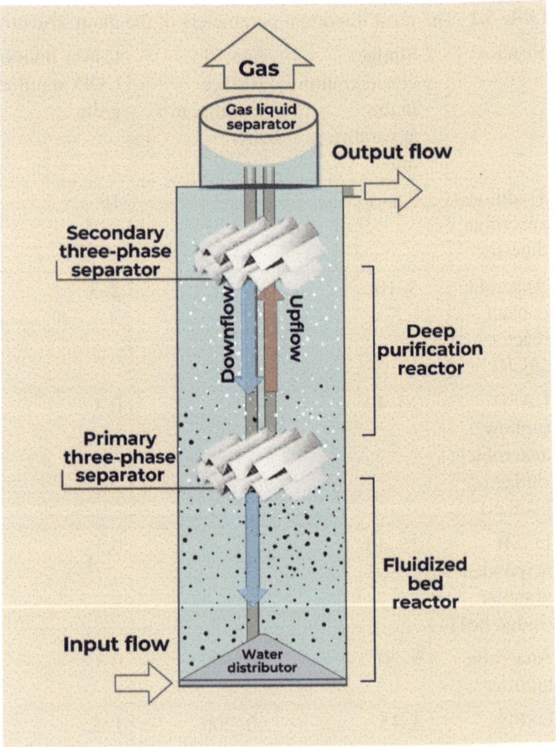

Fig. 5.16 Schematic diagram of internal circulation (IC) reactor

5.3.3 Parameters for Optimizing the Biogas Production Process

It is possible to produce biogas using organic materials. The amount of biogas production in m^3 per 1 ton of raw material is indicated in parentheses: liquid manure, solid compost (20–70); biological waste collected on farms (100–200); secondary (re-grown) raw materials (corn silage, non-food grains); sewage and sewage oil (80 150); old fat (1000); grass and biological waste from cattle slaughter farms (100); brewery and distiller waste (20); wastes from warehouses for storage of fruit and wine and dairy farms (25) [28].

To start the fermentation process, it is enough to provide the following conditions:

1. **Maintenance of anaerobic conditions in the reactor**

Maintaining the optimum temperature is one of the most important factors in the fermentation process. Under natural conditions, biogas formation occurs at temperatures from 0 to 97 °C. However, considering the optimization of the process of processing organic waste to produce biogas and biofertilizers, there are three temperature regimes: psychrophilic, mesophilic and thermophilic.

2. **Availability of nutrients**

The growth and activity of methane bacteria require organic and mineral nutrients in the raw material. In addition to carbon and hydrogen, creating biofertilizers requires a sufficient amount of nitrogen, sulfur, phosphorus, potassium, calcium, magnesium and a number of trace elements—iron, manganese, molybdenum, zinc and others.

3. **Fermentation time**

The optimal fermentation time depends on the reactor's loading dose and the fermentation process's temperature. If the fermentation time is chosen too short, then during the unloading of the sludge, the bacteria are washed out of the reactor faster than they can multiply, and the fermentation process is almost stopped. Excessive exposure to raw materials in the reactor does not meet the objectives of obtaining the largest amount of biogas and biofertilizers for a certain time frame.

In determining the optimal duration of fermentation, use the term "turnaround time reactor". The rotation time reactor is the time during which fresh feedstock loaded into the reactor is processed and unloaded from the reactor.

For systems with continuous loading, the average fermentation time is determined by the reactor volume ratio to the daily volume of loaded feedstock. In practice, the reactor rotation time is chosen depending on the fermentation temperature and the composition of the raw material in the following intervals:

- Psychrophilic temperature: from 30 to 40 days or more.
- Mesophilic temperature: from 10 to 20 days.
- Thermophilic temperature regime: from 5 to 10 days.

The daily dose of feedstock is determined by the reactor's rotation time and increases (as well as the yield of biogas) with increasing temperature in the reactor.

The choice of fermentation time also depends on the type of processed raw materials. For the following types of processed raw materials under conditions of mesophilic temperature, the time during which most of the biogas is released is approximately equal to:

- Liquid cattle manure: 10–15 days.
- Liquid pig manure: 9–12 days.
- Liquid chicken manure: 10–15 days.
- Manure mixed with vegetable waste: 40–80 days.

1. **Acid–base Balance**

Methane-produced bacteria are best suited to live in neutral or slightly alkaline conditions. In the process of methane fermentation, the second stage of biogas production is the phase of active action of acid bacteria. At this time, the pH level decreases, i.e. the environment becomes more acidic.

However, in the normal course of the process, the activity of different groups of bacteria in the reactor is equally efficient, and methane bacteria process acids. The optimal pH value varies depending on the raw material, from 6.5 to 8.5.

2. **The content of carbon and nitrogen**

One of the most important factors influencing methane fermentation (biogas evolution) is the carbon (C) to nitrogen (N) ratio in processed raw materials. If the C:N ratio is excessive, then the lack of nitrogen is a factor limiting the methane fermentation process. If this ratio is too small, then a large amount of ammonia is formed, which becomes toxic to bacteria.

Microorganisms need both nitrogen and carbon to assimilate into their cellular structure. Various experiments have shown that the yield of biogas is the highest at the level of the ratio of carbon and nitrogen from 10 to 20, where the optimum varies depending on the type of raw material. To achieve high biogas production, the mixing of raw materials is practised to achieve the optimal C:N ratio.

3. **The choice of moisture content of raw materials**

Unhindered metabolism in raw materials is a prerequisite for high bacterial activity. This is possible only if the viscosity of the raw material allows the free movement of bacteria and gas bubbles between the liquid and the solids contained in it. There are various solid particles in agricultural waste.

Solid particles, such as sand, clay and others, cause sediment formation. Lighter materials rise to the surface of the raw material and form a crust. This reduces the formation of biogas. Therefore, it is recommended to finely grind plant residues (straw, etc.) and ensure no exogenous particles are in the raw material before loading into the reactor.

4. **Regular stirring**

Periodic stirring is required to efficiently operate the biogas plant and maintain the stability of the raw material fermentation process inside the reactor. The main purposes of mixing are: release of the received biogas; mixing of fresh substrate and bacterial populations (grafting); prevention of crust and sediment formation; diversion of areas of different temperatures inside the reactor; ensuring uniform distribution of the bacterial colonies throughout the entire area of the reactor.

Stirring can be constant or intermittent, depending on the mode of operation of the reactor. The optimal mixing mode significantly reduces raw materials' fermentation time and prevents crust formation on the surface. Although partial mixing occurs due to the release of biogas from the raw material, temperature movement and as a result of fresh raw material, such mixing is not enough.

Stirring should be done regularly. Rarely mixing the raw materials will lead to stratification of the raw material mass and crust formation, thereby reducing the efficiency of gas formation. Well-mixed raw materials can give 50% more biogas. However, stirring too often can damage the enzymatic processes inside the reactor. However, irregular stirring can also result in incomplete raw material processing as the microbial consortium can be disrupted, and the fermentation process will be unproductive until the new consortiums of bacteria are formed. This is due to the process of symbiotic fermentation between different strains of bacteria, i.e. bacteria of one species befit from another. Hence, there is a need to ensure the homogeneity of the inoculated medium, and it is recommended to mix the raw material slowly and consistently every 4–6 h.

5. **Process inhibitors**

Some inorganic substances do not contribute to the function of microorganisms; fermented organic matter should not contain substances that adversely affect the activity of microorganisms (antibiotics, solvents, etc.). These slow down and sometimes stop the process of biogas production. Each of the different types of bacteria involved in the three stages of methane formation is affected differently by these parameters.

There is also a close interdependence between the parameters (for example, the choice of fermentation time depends on the temperature regime), so it is difficult to determine the exact effect of each factor on the amount of biogas produced. Table 5.3 presents a comparative description of anaerobic and aerobic treatment [29, 30].

5.3.4 Calculation of the Main Parameters of Work Biogas Plants

In the technological calculation of biogas plants (methane digester), the following basic parameters must be determined:

- Concentration of dry ash-free matter in the sludge loaded into the methanogenic tank S, kg/m^3.
- Fermentation temperature t, °C.
- Loading on the methanogenic tank based on dry ash-free matter d, kg/(m^3 day).
- Hydraulic retention time of sludge in the methanogenic tank (digestion duration) τ, day.
- Specific biogas yield per unit mass of dry ash-free matter in the loading B_{sp}, m^3/(kg day).
- Biogas yield rate b/τ, m^3/(m^3 day).
- The required working volume of the methane tank V, m^3;
- Total biogas yield B_{total}, m^3/day.

Of great interest for engineering calculations is the Conto model used for the mathematical description of the process of anaerobic fermentation of a number of organic wastes:

$$\overline{\frac{b}{\tau}} = \overline{B} \cdot \left(\frac{S}{\tau}\right)\left(1 - \frac{K}{\mu_m \tau - 1 + K}\right), \tag{5.17}$$

where $\overline{b/\tau}$—methane yield rate, m$^3_{CH4}$/(m^3 day); $\overline{B}$—maximum yield of methane per unit mass of organic matter loaded into the methane tank at infinitely long process

Table 5.3 Comparative analysis of aerobic efficiency and anaerobic reactors

Anaerobic reactor	Aerobic reactor
1. Resistance to high concentrations of organic matter (satisfactorily operates at BOD more than 1000 mg/l)	1. Resistance to high concentrations of organic matter (operates at BOD up to 100 mg/l)
2. Resistance to intermittent supply and composition of wastewater	2. Sensitivity to intermittent supply and composition of wastewater
3. Resistance to nutrient deficiency in wastewater	3. Sensitivity to nutrient deficiency in wastewater
4. Low sludge growth compared to aerobic reactors (10 times less)	4. The problem of excess sludge disposal
5. Low operating costs (energy savings on mixing)	5. High operating costs (requirement for aeration)
6. Biogas production	6. Absent
7. High resistance to toxicants and the ability to recover performance	7. Low resistance to toxicants and low ability to recover performance
8. High energy content of the final product	8. Low energy content of the final product
9. Low process speed	9. High process speed
10. Low biomass growth (during initial startup)	10. High biomass growth (during initial startup)
11. Moderate BOD removal efficiency (60–90%)	11. High BOD removal efficiency (95%)
12. Does not remove nitrogen and phosphorus	12. Removes nitrogen and phosphorus
13. The cost of sealing, heating	13. Absent
14. Explosion hazard	14. Absent
15. Formation of sulfides	15. Absent
16. Low consumption of biologically usable energy	16. High consumption of biologically usable energy
17. No self-heating of sludge (heat supply required)	17. Self-heating of sludge
18. No foaming	18. High foaming with sulfur compounds accumulating in the upper sludge layer, bulking of sludge

duration, $m^3{}_{CH4}/kg$; S—concentration of organic matter in the loaded sludge, kg/m^3; τ—duration of fermentation, day; $K = f(S)$ the kinetic parameter of the process.

The maximum specific growth rate of biomass depends on the fermentation temperature and is calculated by the formula

$$\mu_m = 0.013 \cdot t - 0.129\,\text{day}^{-1}, \tag{5.18}$$

where t is the fermentation temperature, °C.

As shown in Eq. (5.17), the rate of methane yield depends on the concentration of organic matter S in the initial substrate. The concentration of organic (dry ashless) in the sediment, kg/m^3, depends on the humidity W, ash content on dry weight A and the bulk density of the sediment $\rho\rho$:

$$S = \rho(100 - W) \cdot (100 - A) \cdot 10^{-4} \quad (5.19)$$

As noted above, the ash content of the sludge affects the fermentation limit. The anaerobic digestion process of sludge depends on the mass ratio of water (G_w) to organic matter (G_{org}) in the initial sludge. After transformation, the following equation can be derived:

$$G_w / G_{org} = W \cdot \rho_s / S \quad (5.20)$$

The limit of sludge fermentation depends on the chemical composition of the organic substances. It is known that the bulk density of the sediment ρ_s depends on its true density ρ_t, which, in turn, depends on the mineral density ρ_m, organic ρ_{org} parts of the dry matter of the sludge. Due to its flammability, dry sludge can be considered a solid fuel. Therefore, it is fair to determine the density of the organic matter of the sludge as for solid fuels by known methods, according to which ρ_{org} depends on the content in the combustible mass of fuel carbon C^c and hydrogen H^c.

It can be stated that they all affect the fermentation process by assessing the role of three indicators in the formula (5.19)—humidity W, ash content A and bulk density ρ. Therefore, the concentration of organic matter S can be considered as a complex criterion that characterizes the properties of the sludge as a substrate during anaerobic fermentation. Experimental studies show that the value of S affects the duration of fermentation required to reach the technical limit of fermentation, which increases with increasing S. When exceeding a certain (critical) concentration of organic matter S_{kp} (for a number of organic wastes $S_{kp} = 60–100$ kg/m^3) the fermentation process is inhibited, while the value of the kinetic parameter K in Eq. (5.17) increases sharply. It is believed that the increase in K indicates inhibition of the process caused by overloading the system, the presence of inhibited substances in excess of acceptable levels, deterioration of contact between bacteria and organic substrate due to high concentrations of the latter and other reasons.

The technological parameter—simplex b/τ(or $\overline{b/\tau}$)—was not used in the domestic practice of calculations of the anaerobic fermentation process. The parameter used instead, b, is the gas output per unit volume of the methane tank or, almost the same, per unit volume of fermented sludge. This parameter was used to roughly estimate the total biogas yield at treatment plants. However, increasing the rate of biogas output b/τ enables maintenance services to increase the total biogas output without changing the fermentation technology. Another way to regulate the output of biogas, which is to change the volume of methane tanks, is now hardly feasible, as most designs of methane tanks have a constant geometry and volume, and keeping spare methane tanks is not economically feasible. Increasing the load d allows, other things being equal, to pass through the existing apparatus a proportionately larger amount of sediment. Designing new structures means reducing the cost of building smaller methane tanks. At the same time heat losses due to the designs of the protecting methane tanks decrease also. In practice, the real possibilities of increasing the load on the methane tank for sludge of a given concentration are exhausted already during design because, at this stage, the maximum possible conditions of the loading dose, the minimum duration of fermentation and the maximum load on the methane tank are usually assumed. Thus, for a traditional fermentation process, it is only possible to increase the load by increasing the sludge concentration. The use of concentrated precipitation allows not only the reduction of the volume of methane tanks but also, in accordance with the reduction of precipitation, the further reduction of heat consumption for heating to maintain a given fermentation temperature. It follows from Eqs. (5.17) that under a certain load

$$d = S/\tau \tag{5.21}$$

methane release rate b/τ depends on the maximum possible yield of methane $\overline{B}$, residence time of the fermented biomass in the methane tank and kinetic parameters μ_m and K.

Experiments with livestock waste have shown that $\overline{B}$ depends on the breed of animals that secrete manure, their diet, the term and method of manure storage, the amount of foreign materials and other factors that determine the chemical composition of manure. The fermentation temperature does not affect the ultimate methane yield. Since, according to the definition, the following equation are always valid for calculating the specific yield of methane (biogas) per unit mass of organic matter of the initial substrate:

$$\overline{B_{nu\mathrm{T}}} = \frac{\overline{b}}{\tau} : \frac{S}{\tau} \tag{5.22}$$

or

$$\overline{B_{nu\mathrm{T}}} = \frac{\overline{b}}{\tau} : d \tag{5.23}$$

then you can write

$$\overline{B_{nu\mathrm{T}}} = \overline{B}\left(1 - \frac{K}{\mu_m - 1 + K}\right) \tag{5.24}$$

As the duration and temperature of fermentation increase, the specific yield of biogas and, accordingly, the decomposition of organic matter increases.

Size $\overline{b/\tau}$ has a clear extreme. Value $\overline{b/\tau}$ and d at the point of bending, which we denote by the index (t). Calculations show that

$$\overline{b/\tau}_{(m)} = \overline{B}\frac{\mu_m S}{\left(1 + \sqrt{K}\right)^2} \tag{5.25}$$

$$d_{(m)} = \mu_m S/\left(1 + \sqrt{K}\right) \tag{5.26}$$

$$\overline{B_{nuT\,(m)}} = \frac{\overline{B}}{\left(1 + \sqrt{K}\right)} \tag{5.27}$$

The highest methane release rate is achieved at $S = S_{cr}$ because, in this case, $K = 1$. Then the formulas (5.25)–(5.27) will look like this:

$$\overline{b/\tau_{(m)}} = \overline{B}\left(\mu_m S_{\mathrm{Kp}}/4\right) \tag{5.28}$$

$$d_{(m)} = \mu_m S_{\mathrm{Kp}}/2 \tag{5.29}$$

$$\overline{\mathrm{B}_{nuT\,(m)}} = \frac{\overline{B}}{2} \tag{5.30}$$

Unfortunately, during the fermentation of liquid precipitation of urban wastewater by traditional technology, limit loads $d_{(m)}$ usually cannot be used due to disruption of the normal process, which is expressed in an increase in the formation of volatile fatty acids, a decrease in the proportion of methane and an increase in the proportion of carbon monoxide in biogas. Under the mesophilic fermentation regime, the load on the methane tank is usually kg/(m^3 day): in France—0.8 to 1.2; in England—1.0 to 2.0; in Germany—2.5 to 4.0; in the USA—2.5 to 5.0.

According to the method of AKG, given in SNiP 2.04.03–85, usually, the calculation of the decomposition of dry ashless sediment liquid loaded into the metatank is carried out according to the formula

$$R_r = \mathrm{R}lim - K_r D_{mt}. \tag{5.31}$$

Due to the correlation between the decomposition of dry ashless matter R_r and specific biogas yield B_{sp} at a constant density of biogas $\rho_r (R_r = \mathrm{B}_{sp}\rho_r \cdot 100\%)$ and assuming a constant proportion of methane in the biogas, we can show that the value of B_{sp} during the calculations according to the specified state standard and the kinetic model of Conto for the case of $S \leq S_{cr}$ determined by a single case:

$$\mathrm{B}_{sp} = \mathrm{B} - K_r{}^{\tau - 1} \tag{5.32}$$

where B is the maximum yield of biogas per unit mass loaded into the methane tank of organic matter for an infinite duration of fermentation, m^3/kg; τ—duration of fermentation, day; K_r is the coefficient of proportionality, m^3 day/kg, determined by the function

$$Kr = BKS/[\mu_m S - d(l - K)] \tag{5.33}$$

For precipitation of urban wastewater with a humidity of 93–97%, as shown by experimental data, the decay limit can be taken as constant. According to specified state standard in this range of humidity in the absence of data on the chemical composition of precipitation, the maximum yield of biogas B, m^3/kg of dry ashless substance, can be taken as equal to 0.53 for the sediment of primary settling tanks and 0.44—for excessively active sludge. In this case, the coefficient of proportionality can be calculated by the empirical formula

$$K_r = \frac{(38S - 205)P}{100(t - 17.8)} \tag{5.34}$$

where t is the fermentation temperature, °C; P is the correction factor.

Formula (5.34) can be used to calculate the values of the critical precipitation concentration S_{cr}:

$$S = [100K_p(t - 17.8) + 205]/(38P) \tag{5.35}$$

Given that for $K = 1$ Eq. (5.33) will take the form $K_r = B/\mu_m$, we obtain

$$S_{cr} = [100B - 1(t - 17.8) + 205]/(38P) \tag{5.36}$$

The presence of toxic substances and a decrease in the amount of easily soluble organic substances leads to an additional increase in the coefficient $K_{r,}$ which is determined from experimental data.

The value of the critical concentration of precipitation S_{cr}, calculated for mesophilic and thermophilic modes of anaerobic fermentation, are shown in Table 5.4.

Calculations according to the specified state standard method and according to the kinetic model of Konto at $S \leq S_{kr}$ obtain the same results if we determine the coefficient K_r by formula (5.36) or empirical formula (5.37) or, conversely, calculate the value of the kinetic parameter K by the equation derived from formula (5.34):

Table 5.4 Values of the critical concentration of dry ashless (organic) substance S_{cr}

Sediment type	Fermentation limit, %	The value of S_{cr}, kg/m^3 According to the fermentation regime	
		Mesophilic (33 °C)	Thermophilic (53 °C)
Precipitate from primary settling tanks	53	76	93
Excess activated sludge	44	64	78

Note that in contrast to the Conto data obtained mainly during manure fermentation, where $S \leq S_{cr}$ value $K = 1$, during the fermentation of liquid precipitation of urban wastewater $K < 1$. For example, during mesophilic fermentation of raw sludge from primary settling tanks with a humidity of 93–95%, the values of K calculated on the basis of experimental data are 0.5–0.3, respectively

$$K = Kr(\mu_m S - d)/(BS - Krd) \tag{5.37}$$

The size of the reactor is measured in cubic meters and depends on the quantity, quality and type of raw material, as well as on the selected temperature and fermentation time. There are several ways to determine the required reactor volume.

The ratio of the daily feedstock dose and the reactor size. The daily dose of feedstock is determined based on the fermentation time (reactor rotation time) and the selected temperature. For the mesophilic fermentation regime, the reactor rotation time is from 10 to 20 days, and the daily loading dose is from 1/20 to 1/10 of the total volume of raw materials in the reactor (Table 5.5). The initial loading of the reactor with the substrate is carried out on 2/3 of the volume.

Continuous loading and unloading

Methane tanks can operate in batch, continuous and semi-continuous modes. The transition to a continuous fermentation mode while maintaining the value of the decay leads to some increase in the volume of the reactor.

The size of the reactor for processing a certain amount of raw materials

First, based on the number of animals, experimentally determine the daily dose of manure (DD) for processing in a biogas plant. Then, the raw material is diluted with water to achieve 86–92% moisture. In most rural plants, the ratio of manure to water mixed to produce raw materials ranges from 1:3 to 2:1. Thus, the amount of loaded raw materials (DL) is the sum of farm waste (DD) and water (DW), which are diluted. For processing raw materials under the mesophilic mode, it is recommended to use a dose of daily loading DL equal to 10 t of the total volume of the substrate loaded in installation (TSV). The total volume of raw materials in the reactor should not exceed 2/3 of the reactor volume. Thus, the volume of the reactor (RV) is calculated by the following formula:

$$\text{TSV} = 2/3 \cdot \text{RV}, \text{ and RV} = 1.5 \cdot \text{TSV} \tag{5.38}$$

where

$$\text{TSV} = 10 \cdot \text{DL} \tag{5.39}$$

Table 5.5 Reactor loading dose during liquid-phase fermentation

Humidity	Load dose at different temperature regimes, %	
	mesophilic (33 °C)	thermophilic (53 °C)
93	7	14
94	8	16
95	9	18
96	10	20
97	11	22

$$\mathrm{DL} = \mathrm{DD} + \mathrm{DW} \tag{5.40}$$

If the homestead contains 10 heads of cattle, 20 pigs and 35 chickens, the volume of daily excrement from 1 cattle = 55 kg, from one pig = 4.5 kg, from 1 chicken = 0.17 kg. The volume of daily waste of the DD farm will be approximately equal to 646 kg. Cattle and pig excrement humidity will be 86%, and chicken manure will be 75%. To achieve 85% humidity, 3.9 l of water (approximately 4 kg) must be added to the bird droppings. Therefore, the daily dose of raw material loading will be approximately 650 kg. The full load of the reactor TSV = 10 × 0.65 = 6.5 tons, and the volume of the reactor RV = 1.5 × 6.5 = 9.75, or approximately 10 m^3.

To calculate the concentration of dry ashless sludge S is necessary to determine the bulk density of the sediment ρ_{bulk} a given humidity W. The value ρ_{bulk} be calculated by the formula

$$\rho_{\text{bulk}} = \frac{\rho_t}{100 + W\left(\rho_t \cdot 10^{-3} - 1\right)} \cdot 100 \tag{5.41}$$

where ρ_t is the true sediment density (solid phase density); $\frac{\text{kg}}{\text{m}^3}$; the density of the dispersion medium (water) precipitation is taken as 1000 $\frac{\text{kg}}{\text{m}^3}$.

The specific yield of biogas can be calculated by formula (5.32) or the formula that follows from it:

$$\mathrm{B}_s = \mathrm{B} - \mathrm{K}_g(d/S) \tag{5.42}$$

and when $S > S_{cr}$—according to the modified formula (5.24) considering the assumption of constancy of the methane content in biogas:

$$\mathrm{B}_s = \mathrm{B} - \left(1 - \frac{\mathrm{K}}{\mu_m - 1 + \mathrm{K}}\right) \tag{5.43}$$

or

$$\mathrm{B}_g = \mathrm{B}\frac{\mu_m{}^{S-d}}{\mu_m{}^{S+d(\mathrm{K}-1)}} \tag{5.44}$$

where B—the maximum yield of biogas per unit mass of dry ashless substance in the sediment loaded into the m^3/kg methane tank.

Decomposition of organic matter R_g,%, calculated by the formula

$$R_g = \mathrm{B}_s \rho_g \cdot 100 \tag{5.45}$$

where ρ_r ρ_r is the density of biogas kg/m^3.

The rate of biogas yield, m^3/(m^3 day), is determined by the equation

$$b/\tau = B_s d \tag{5.46}$$

The required working volume of the methane tank V, m^3 for a given amount of fermented sludge Q, m^3/day can be determined by the formula

$$V = Q\tau \tag{5.47}$$

The total biogas yield, m^3/day, is calculated by the formula

$$B_{tot} = \left(\frac{b}{\tau}\right)V \tag{5.48}$$

Further calculation of methane tanks is reduced to determining the unit volume and shape of methane tanks, their thermal calculation, definition of the technological scheme (biogas utilization), calculation of biogas and heat used equipment, and calculation of pumps, mixers and other ancillary equipment.

5.4 Bio Purification of Gas and Air Emissions

Biological methods of air purification are based on the ability of microorganisms to destroy, under aerobic conditions, a wide range of substances and compounds to the final products—CO_2 and H_2O. The ability of microorganisms to metabolize aliphatic, aromatic, heterocyclic, acyclic and various C_1 compounds is widely known. Microorganisms utilize ammonia and oxidize sulfur dioxide, hydrogen sulfide and dimethyl sulfoxide. The formed sulfates were utilized by other microbial species. There is evidence of effective oxidation by aerobic carboxybacteria of carbon monoxide, which is one of the most dangerous air pollutants. Representatives of the genus *Nocardia* effectively destroy sterols and xylene; *Hyphomicrobium*—dichloroethane; *Xanthobacterium*—ethane and dichloroethane; *Mycobacterium*—vinyl chloride. Three types of installations are used for biological air purification: biofilters, bioscrubbers and bioreactors with a washed layer [31] (Table 5.6).

In biofilters, for a filter carrier for a layer, use natural materials—compost, peat, etc. These materials contain various mineral salts and substances necessary for developing microorganisms. Therefore, biofilters do not make any mineral additives. The air to be cleaned is supplied by a fan to the system and passes through the filter layer in any direction, from bottom to top or vice versa. The air must pass evenly through the entire volume of the filter layer. Therefore, the homogeneity of the layer and a certain degree of humidity are required.

The optimum humidity of the filter layer for air purification is 40–60 tons by weight of the carrier material. In case of insufficient moisture in the material of the filter layer, cracks are formed in it and the material dries out.

The gas-dynamic parameters of the filter layer, the spectrum and concentration of substances present in the air and the enzymatic activity of destructive microorganisms

Table 5.6 Classification of biological air treatment plants

Type of system	Working medium	Water regime	Main stages of air pollutant removal	Source of mineral salts
Biofilter [32]	Filtration layer with immobilized microbial cells on natural or synthetic carriers	No water circulation observed	1. Desorption by the filtration layer material 2. Degradation by microbial cells	Filtration layer with immobilized microbial cells on natural or synthetic carriers
Bioscrubber [33]	Water, activated sludge	Water circulation	1. Absorption in the water absorber 2. Degradation in the aeration tank with activated sludge	Water, activated sludge
Trickling Biofilm Reactor [34]	Immobilized microbial cells on artificial carriers	Water circulation	1. Diffusion through the water film to microorganisms 2. Degradation in the biological layer	Immobilized microbial cells on artificial carriers

determine the efficiency of the biofilter. The rate of removal of harmful impurities from the air in the process of biotreating can be limited by the diffusion of substances from the gas phase into the biocatalytic layer and the rate of biochemical reactions in microbial cells [35].

At a high input concentration of harmful substances in the air, their destruction in the flow through the filter layer is uneven. Initially, readily available substances are destroyed, and only at the end of the process does the destruction of difficult-to-degrade compounds begin. Thus, in the presence of a complex of compounds (butanol, ethyl acetate, butyl acetate and toluene) in the air as harmful impurities, the latter was utilized by microorganisms only after oxidation of all other substances.

The principle of operation of bioscrubbers differs in that the air purification process is realized in two stages in two different units. In the first stage, in the absorber, toxic substances contained in the air and oxygen are soluble in water. As a result, the air is purified, and the contaminated water is further purified. Various types of absorbers are used, including bubble (barbotage) absorbers, packed bed absorbers, spray absorbers, nozzle absorbers, and others. The purpose of design improvements is to increase the surface area of the phase separation, gas liquid. This determines the efficiency of absorption. In the second stage, the contaminated water enters the aeration tank, where it is regenerated. Water purification in the aeration tank occurs according to the usual scheme with the participation of oxygen. In the purification process, complex organic substances are oxidized by microorganisms that form activated sludge to the final products with biomass formation.

Compared to biofilters, bioscrubbers take up less space because they are towers several meters high. Operating costs when using bioscrubbers are higher because the process of bio-purification of water requires significant costs. The use of bioscrubbers is effective in the presence of well-soluble toxic substances in the air.

Biological methods of gas and air emissions purification have started to be used relatively recently, but so far, they have been on a limited scale. However, the prospect of their use is evident.

5.5 Bioleaching of Useful Components from Mineral Raw Materials

Bacterial leaching means the selective extraction of chemical elements from multi-component compounds in the dissolution process of microorganisms from an aqueous medium. Bacterial leaching can be used in any leaching method if temperature and pressure are not used [36].

Bioleaching is a technology that allows the extraction of metals from depleted ores by using chemolitotrophic microorganisms.

Bioleaching saves materials and energy; it can replace traditional methods of mineral processing such as roasting, autoclave leaching, and metallurgical smelting, which pollute the environment with toxic gases and toxic chemicals. These bacteria are harmless to humans, feed on minerals, are easily transported, are resistant to low temperatures and lack of nutrient medium and can exist at temperatures up to 80 °C [37].

Table 5.7 lists the microorganisms important for hydrometallurgy and possible areas of their application [38, 39].

Biochemical leaching is performed by aerobic bacteria and archaea, which can oxidize sulfide minerals. Leading in these processes are representatives of the genera Acidithiobacillus, Leptospirillum, Sulfobacillus, Sulfolobus, Acidianus, Metallosphaera, and Ferroplasma. In direct bacterial leaching, microbial cells interact with the surface of the solid particle. In recent decades, bioleaching research has been developing the use of microorganisms that conduct leaching processes at elevated temperatures capable of destroying silicate minerals and extracting metals at moderately acidic pH values.

During research in the cultivation process, a stable association of aerobic microorganisms is formed, reaching at least 4.9×10^{10} CFU/ml.

Thus, it was investigated that on the substrate containing fresh phosphogypsum, it is possible to achieve high development rates of the necessary ecological and trophic groups of bacteria involved in the biochemical leaching of useful elements from phosphogypsum. From the perspective of the use of phosphogypsum as a mineral substrate in biotechnological processes, the following advantages can be identified: low cost; is a source of necessary macro-and microelements for microorganisms;

Table 5.7 Microorganisms are important for hydrometallurgy

Microorganisms	Processes	Possible scope
Bacteria of the genera *Thiobacillus* and *Leptospirillum* (*Thiobacillus ferroxidans, Thiobacillus thioxidans, Thiobacillus acidophilus, Leptospirillum ferroxidans*)	Oxidation of sulfide minerals S° and Fe^{2+} at pH 1.4–3.5 and t = 5–35 °C	Underground and vat leaching of metals from sulfide and mixed ores and concentrates, from wastes of the pyrometallurgical industry
Facultative thermophilic bacteria are close to thiobacilli	The same at pH 1.1–3.5 and t = 30–55 °C	The same
Facultative thermophilic bacteria of the genus *Sulfobacillus*	The same at pH 1.1–4.0 and t = 28–60 °C	The same
Acidophilic bacteria of the genera *Sulfolobus and Acidianus*	The same at pH 1.0–5.0 and t = 45–96 °C	The same
Heterotrophic microorganisms and their metabolites (bacteria, micelles, yeasts, algae)	Destruction of sulfide, silicate, aluminosilicate minerals, reduction and oxidation of manganese, dissolution of gold, bisorption of metals	Extraction of metals from carbonate and silicate ores and rocks, ore beneficiation; leaching of gold; use of bacterial biomass and metabolites in flotation of ores (lipids); treatment of industrial wastewater from metals; selective extraction of metals from solutions

stimulates the development of the necessary ecological and trophic groups; creates favourable conditions for the formation of a stable biofilm.

Bioleaching saves materials and energy; it can replace methods of mineral processing such as firing, autoclave leaching, and metallurgical smelting, which pollute the environment with toxic gases and chemicals. These bacteria are harmless to humans, feed on minerals, are easily transported, are resistant to low temperatures and lack of nutrient medium, and can exist at temperatures up to 80 °C.

5.5.1 *The Method of Surface Leaching of Piles and Dumps*

This method is usually used to leach copper from rocks with low content (less than 0.4% by weight). Such dumps accumulate in large quantities during large-scale open pit mining of ore and can occupy huge areas and reach a height of several hundred meters. Leaching piles are somewhat different from leaching dumps. The piles contain a high content of metals compared with the dumps, the extraction of which is, in principle, possible in a relatively short time—a few months. At the same time, the leaching of dumps can take years [40].

In piles and dumps, crushed ore is placed on a sloping waterproof stand. The surfaces of piles and dumps are irrigated with a liquid, which is a weak solution of acid and ferrous ions. The solution with the leached metal, filtered through a layer of rock, is collected from below. As natural microorganisms usually develop during the leaching of dumps in the environment, sowing (inoculation) is not performed. An acidic environment and the presence of oxygen increase the catalytic activity *of Thiobacillus ferrooxidans*. The liquid is leached with the help of pumps spraying it on the top surface of the pile or dump ore, and then it flows down by gravity and is filtered through this ore. Metal-enriched solutions flowing from dumps and piles are then sent to special ponds and reservoirs to collect and extract metal. Extraction is carried out by simple precipitation or electrolysis, etc. The spent solution, containing mainly dissolved iron, is regenerated in oxidizing ponds and re-fed to the dumps [41].

A typical scheme of bacterial leaching of copper from piles and dumps is presented in Fig. 5.17.

The metal extraction rate in the industrial leaching of piles and dumps depends on many factors: the activity of the culture, the quality of the ore and the level of its dispersion, the filtration rate of the leaching solution, and aeration. Thus, with the introduction of compressed air into the layer of leached copper ore, the copper extraction rate increases by 25%.

Less often, microorganisms are used for leaching uranium on an industrial scale. To do this, the rock or ore must be rich in sulfide minerals and not too intensely absorb oxygen.

Figure 5.18 shows the scheme of heap leaching, which is possible for applying metal extraction from ore material.

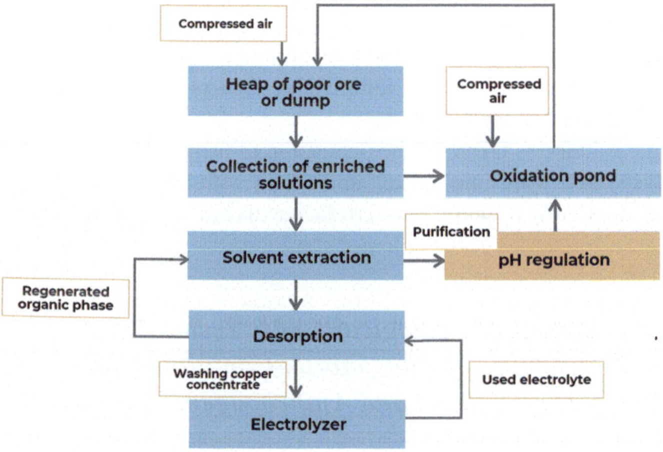

Fig. 5.17 Scheme of bacterial leaching of copper from piles or dumps of ore

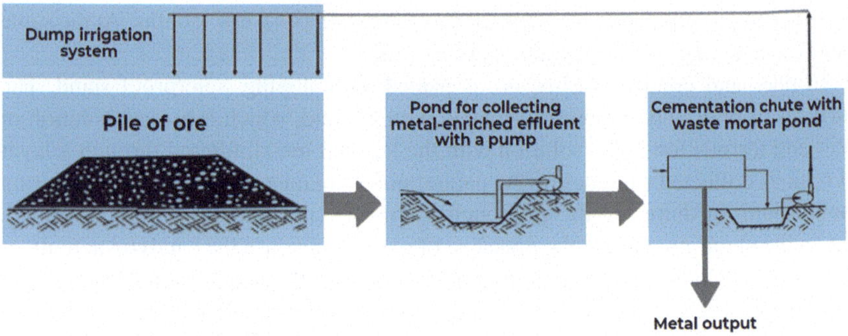

Fig. 5.18 Scheme of the process of heap leaching

A pile of crushed material, classified into specific fractions, is placed on a mesh substrate to allow the free passage of the biosuspension, which is filtered through a layer of phosphogypsum. The biosuspension or chemical solution is fed by pump through the irrigation system. Beneath the mesh, on the ground, collection channels are arranged to direct the productive solution (bio-suspension with extracted metal) into a collection pond, from where it is pumped into cementation channels for metal separation from the pulp.

The ores are crushed to a size class of 100% minus 7.5 m and stacked on a waterproofing base in stacks 5 m high, weighing 1000 tons. There are several stacks in operation. Ore stacks are irrigated with acidic solutions (pH 1.4–1.9), with bacterial oxidation of sulfides, and iron and non-ferrous metals pass into the solution. The solutions are in circulation, and before irrigating the ore stack, the solutions are passed through a stack of inert material (hollow rock). In this stack, bacterial oxidation of dissolved ferrous iron (Fe^{2+}) to ferric iron (Fe^{3+}) occurs. Thus, solvent regeneration is achieved.

Due to the oxidation of sulfides, the temperature in the stack reaches 90 °C. The optimum temperature is 40–50 °C. Temperature control is carried out by changing the air flow and solution for irrigation.

Part of the solution is taken for the deposition of metals. First, copper is isolated by electrolysis, iron with lime, nickel, and cobalt with magnesium carbonate. The process lasts about a year, while non-ferrous metals are extracted at 70–90%.

5.5.2 Development of Ecologically Safe Technologies for Bioleaching of Phosphorus-Containing Raw Materials of Natural and Anthropogenic Coordination

Selective reduction of low-quality phosphates can be carried out in two stages. The first stage is based on the “principle of bioleaching”, which is widely used to extract metals (eg, copper, zinc, uranium, etc.) in the mining industry.

The fundamental technological scheme of the biooxidation process of phosphorus-containing raw materials is shown in Fig. 5.19.

By means of microbial generation, sulfuric acid, phosphate derivatives, and heavy metals can be dissolved within a few hours. The remaining solid phase is separated from the liquid phase and can be disposed of. In the second stage, phosphate is selectively separated from heavy metals (biosorption). Then, phosphate-enriched biomass is separated from the liquid phase and can be returned to the process. Thus, this process has significant advantages (Fig. 5.20) and allows for restoring up to 90% of the original phosphate [42].

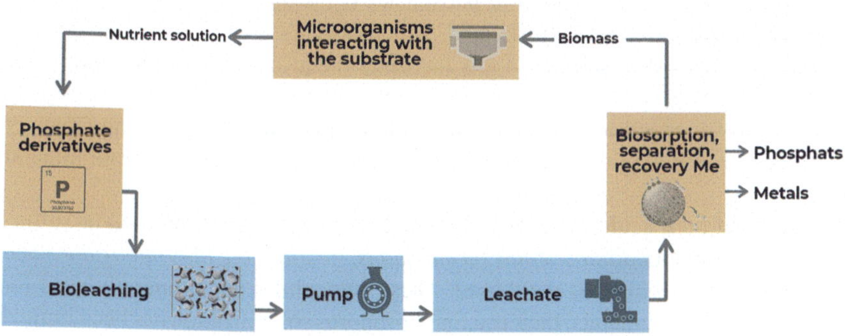

Fig. 5.19 Technological scheme of biooxidation process

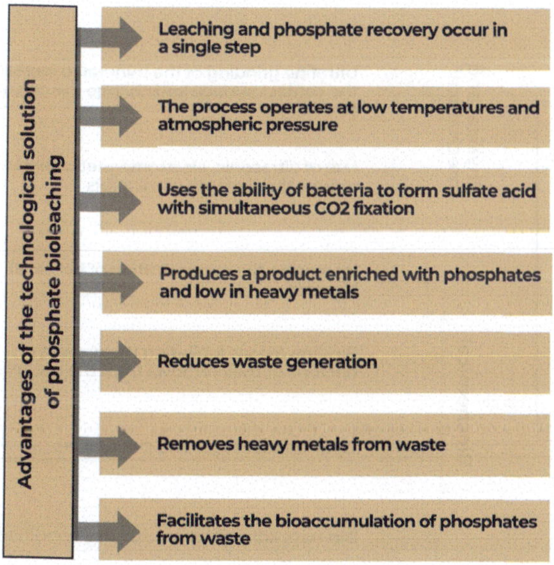

Fig. 5.20 The main ecological and biochemical aspects of processing phosphate raw materials in the process of biochemical conversion

Phosphorus compounds, which are part of minerals, are inaccessible or poorly accessible to plants. However, many microorganisms can convert insoluble phosphoric acid compounds into a soluble state. These include representatives of bacteria, actinomycetes, fungi and other groups of microorganisms (genera *Pseudomonas. Bacillus, Micrococcus, Mycobacterium, Penicillium, Aspergillus*). Dissolution of phosphates in the soil occurs due to the formation of carbon dioxide or various acids [43].

The mobilization of insoluble phosphorus compounds also occurs due to the formation by microorganisms of organic acids and keto acids during incomplete oxidation of carbohydrates or their fermentation.

In some cases, the dissolution of phosphates is facilitated by nitric acid formed by nitrifying bacteria and sulfuric acid, which appears as a result of the activity of bacteria that oxidize sulfur. All this increases the availability of phosphorus for plants.

In industry, the methods shown in Fig. 5.21 are used quite effectively to intensify the processes of biological oxidation or reduction of useful elements from mineral ores.

Mixed and pure acidophilic bacterial cultures consisting of iron- and/or sulfur-oxidizing bacteria *Acidithiobacillus ferrooxidans, Acidithiobacillus thiooxidans* and *Leptospirillum ferrooxidans* were used. These acidophiles are commonly used for bioleaching sulfide minerals; their use for phosphorus solubilization has been limited thus far. The yield of phosphorus bioleaching was up to 97% and 28% for fluorapatite ore of low concentrate and concentrate, respectively, in solutions with a solids content of 1% [44].

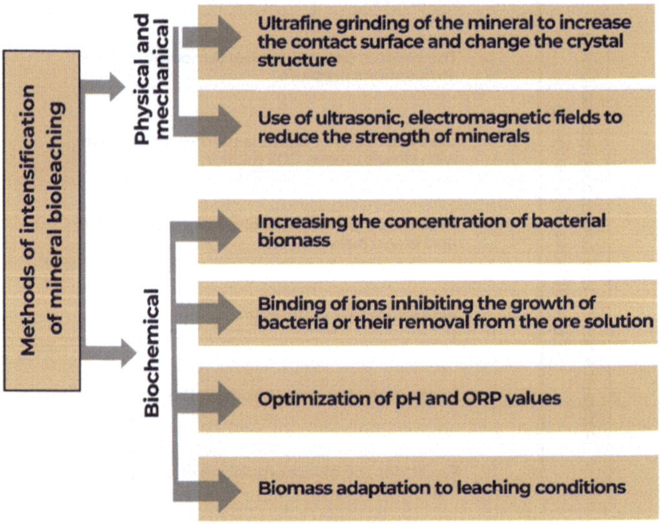

Fig. 5.21 Methods of intensification of bacterial oxidation and leaching of minerals

It should be noted that the natural mechanisms of autocatalysis in the bioreactor space form its association of microorganisms that transform the raw material, and there is a high probability that the dominant species in this association will not be artificially inoculated strain, and natural species adapted to changed conditions. nutrient medium. This phenomenon may be justified by synergetic patterns of microecosystem development in the bioreactor space [45].

Accordingly, the priority is not to isolate highly active strains but to study the synergetic patterns of the dominance of species of microorganisms involved in the biological conversion of minerals. It may be present in symbiotic relationships in the microbial association (which may contain both heterotrophic and chemolithotrophic groups). The industrial process of environmentally friendly production of phosphate fertilizers and raw materials with low phosphorus content aims to incorporate the biodegradation of raw phosphate materials and organic matter under acidophilic conditions.

References

1. Exploring Engineering (2021) Bioengineering. In: Exploring engineering, 5th edn. https://www.sciencedirect.com/topics/engineering/bioengineering
2. Cytotherapy (2011) Bioreactor. Immunol Microbiol. https://www.sciencedirect.com/topics/immunology-and-microbiology/bioreactor
3. Gresch M, Armbruster M, Braun D, Gujer W (2011) Effects of aeration patterns on the flow field in wastewater aeration tanks. Water Res 45(2):810–818. https://doi.org/10.1016/j.watres.2010.09.009
4. Fayolle Y, Cockx A, Gillot S, Roustan M, Héduit A (2007) Oxygen transfer prediction in aeration tanks using CFD. Chem Eng Sci 62(24):7163–7171. https://doi.org/10.1016/j.ces.2007.08.082
5. Qambar AS, Al Khalidy MM (2022) Optimizing dissolved oxygen requirement and energy consumption in wastewater treatment plant aeration tanks using machine learning. J Water Process Eng 50:103237. https://doi.org/10.1016/j.jwpe.2022.103237
6. Habermacher J, Benetti AD, Derlon N, Morgenroth E (2015) The effect of different aeration conditions in activated sludge—side-stream system on sludge production, sludge degradation rates, active biomass and extracellular polymeric substances. Water Res 85:46–56. https://doi.org/10.1016/j.watres.2015.08.002
7. Valchev D, Ribarova I, Borisov B et al (2024) Valuable elements in sludge from eight municipal wastewater treatment plants in relation to their recovery potential. Environ Sci Europe 36(11). https://doi.org/10.1186/s12302-023-00837-x
8. Kathi S, Singh S, Yadav R, Singh AN, Mahmoud AED (2023) Wastewater and sludge valorisation: a novel approach for treatment and resource recovery to achieve circular economy concept. Front Chem Eng 5. https://doi.org/10.3389/fceng.2023.1129783
9. Kučera T, Hanušová V (2018) Recirculation of sludge-water in the water treatment process—a pilot study. Water Pract Technol 13(3):461–468. https://doi.org/10.2166/wpt.2018.059
10. Oceans and Human Health (2023) Kyoto protocol. Social Sci. https://www.sciencedirect.com/topics/social-sciences/kyoto-protocol
11. Afotey B, Sarpong GT (2023) Estimation of biogas production potential and greenhouse gas emissions reduction for sustainable energy management using intelligent computing technique. Meas Sens 25:100650. https://doi.org/10.1016/j.measen.2022.100650
12. Danylov DV, Chubur VS, Chernysh YY, Yakhnenko OM (2020) Bioenergy waste recycling: modelling of development trends. Man and Environment. Issues of Neoecology 34:141–152

13. Wu B, Zhang X, Shang D, Bao D, Zhang S, Zheng T (2016) Energetic-environmental-economic assessment of the biogas system with three utilization pathways: combined heat and power, biomethane and fuel cell. Biores Technol 214:722–728. https://doi.org/10.1016/j.biortech.2016.05.026
14. Khan IU, Othman MHD, Hashim H, Matsuura T, Ismail AF, Rezaei-DashtArzhandi M, Wan Azelee I (2017) Biogas as a renewable energy fuel – A review of biogas upgrading, utilisation and storage. Energy Conversion and Management 150:277–294. https://doi.org/10.1016/j.enconman.2017.08.035
15. Banja M, Jégard M, Motola V, Sikkema R (2019) Support for biogas in the EU electricity sector—a comparative analysis. Biomass Bioenerg 128:105313. https://doi.org/10.1016/j.biombioe.2019.105313
16. Achinas S, Achinas V, Euverink GJW (2017) Technological overview of biogas production from biowaste. Engineering 3(3):299–307. https://doi.org/10.1016/J.ENG.2017.03.002
17. Ferdeş M, Dincă MN, Moiceanu G, Zăbavă BŞ, Paraschiv G (2020) Microorganisms and enzymes used in the biological pretreatment of the substrate to enhance biogas production: a review. Sustainability 12(17):7205. https://doi.org/10.3390/su12177205
18. Paolini V, Petracchini F, Segreto M, Tomassetti L, Naja N, Cecinato A (2018) Environmental impact of biogas: a short review of current knowledge. J Environ Sci Health 53(10):899–906. https://doi.org/10.1080/10934529.2018.1459076
19. ScienceDirect (2019) Fluidized bed reactor. In: Biohydrogen, 2nd edn. https://www.sciencedirect.com/topics/chemistry/fluidized-bed-reactor
20. Lim SJ (2006) Comparisons between the UASB and the EGSB reactor. Iowa State University. https://home.engineering.iastate.edu/seungjoo
21. Goddek S, Delaide BPL, Joyce A, Wuertz S, Jijakli MH, Gross A, Eding EH, Bläser I, Reuter M, Keizer LCP, Morgenstern R, Körner O, Verreth J, Keesman KJ (2018) Nutrient mineralization and organic matter reduction performance of RAS-based sludge in sequential UASB-EGSB reactors. Aquacult Eng 83:10–19. https://doi.org/10.1016/j.aquaeng.2018.07.003
22. Lu Q, Jeong B-G, Lai S, Yan Z, Xiao X, Jiang W (2022) Performance comparison of EGSB and IC reactors for treating high-salt fatty acid organic production wastewater. Processes 10:1295. https://doi.org/10.3390/pr10071295
23. Perez KC, Geest JV, Versprille AI, Otten M, Heffernan B (2012) Performance evaluation of a new (Biobed Advanced) EGSB settler
24. Le Tallec X, Vidal A, Thornberg D (1999) Upflow biological filter: modeling and simulation of filtration. Water Sci Technol 39(4):79–84. https://doi.org/10.1016/S0273-1223(99)00059-1
25. Piera M, Lafuente Mazuecos A, Abánades Velasco A, Martínez-Val Peñalosa JM (2010) Hybrid reactors: nuclear breeding or energy production? In: 14th international conference on emerging nuclear systems (ICENES 2009), 29 June–3 July 2010, Ericeira. https://oa.upm.es/8364/2/INVE_MEM_2010_81649.pdf
26. Paques (2024) BIOPAQ® IC: cost-effective anaerobic effluent treatment system for industrial wastewater. https://www.paquesglobal.com/products/biopaq-ic
27. Tran N-H, Bui M-H (2019) Applying internal circulation anaerobic reactor for wastewater treatment: a case study in Saigon paper mill wastewater treatment plant. Arch Civil Eng Environ 12(3):131–142. https://doi.org/10.21307/acee-2019-045
28. Otieno EO, Kiplimo R, Mutwiwa U (2023) Optimization of anaerobic digestion parameters for biogas production from pineapple wastes co-digested with livestock wastes. Heliyon 9(3):e14041. https://doi.org/10.1016/j.heliyon.2023.e14041
29. Ibro MK, Ancha VR, Lemma DB (2024) Biogas production optimization in the anaerobic codigestion process: a critical review on process parameters modeling and simulation tools. J Chem 2024(1):4599371. https://doi.org/10.1155/2024/4599371
30. Habib SS, Torii S, Kavitha M, Nair ACA (2024) Optimization of the factors affecting biogas production using the Taguchi design of experiment method. Biomass 4(3):687–703. https://doi.org/10.3390/biomass4030038
31. Sun P, Yang X, Huang R, Bin H, Zheng S, Wei Z, Xu X, Lu J, Sun X, Chen L (2004) Biopurification of industrial waste gas from rubber regeneration. Waste Manage Res 22(6):492–496. https://doi.org/10.1177/0734242X04048730

32. ECOTEC (2024) Biofilters for biological treatment of gases. https://www.ecotec.es/en/biofilters-for-biological-treatment-of-gas.php
33. Odour Observatory (2024) Bioscrubbing. https://odourobsitory.org/bioscrubbing-2/
34. Reij MW, Keurentjes JTF, Hartmans S (1998) Membrane bioreactors for waste gas treatment. J Biotechnol 59(3):155–167. https://doi.org/10.1016/S0168-1656(97)00169-7
35. Hassan KF, Al-Hussieny AA, Al Deen Hassan ES, Almunim RA (2022) Studying the efficiency of the multiple biofilters in the reduction of pollutants from wastewater. Revista Bionatura 7(2):55. https://doi.org/10.21931/RB/2022.07.02.55
36. Tezyapar Kara I, Kremser K, Wagland ST et al (2023) Bioleaching metal-bearing wastes and by-products for resource recovery: a review. Environ Chem Lett 21:3329–3350. https://doi.org/10.1007/s10311-023-01611-4
37. Castro L, Zhang R, Muñoz JA, Sand W (2024) Editorial: bioleaching and biorecovery of critical raw materials from secondary sources. Front Microbiol 15:1395820. https://doi.org/10.3389/fmicb.2024.1395820
38. Kaksonen AH, Mudunuru BM, Hackl R (2014) The role of microorganisms in gold processing and recovery—a review. Hydrometallurgy 142:70–83. https://doi.org/10.1016/j.hydromet.2013.11.008
39. Cheru MS (2021) Bio hydrometallurgical technology, application and process enhancement. In: Heavy metals—their environmental impacts and mitigation. https://doi.org/10.5772/intechopen.94206
40. Fey DL, Desborough GA (2004) Metal leaching in mine-waste materials and two schemes for classification of potential environmental effects of mine-waste piles. In: Nimick DA, Church SE, Finger SE (eds) Integrated investigations of environmental effects of historical mining in the basin and boulder Mining Districts, Boulder River Watershed, Jefferson County, Montana (professional paper 1652-D4). U.S. Geological Survey. https://pubs.usgs.gov/pp/2004/1652/pdf/508/ChapD4508.pdf
41. Nagaoka T, Ohmura N, Saiki H (1999) A novel mineral flotation process using *Thiobacillus ferrooxidans*. Appl Environ Microbiol 65(8):3588–3593. https://doi.org/10.1128/AEM.65.8.3588-3593.1999
42. Abdelgalil SA, Kaddah MMY, Duab MEA et al (2022) A sustainable and effective bioprocessing approach for improvement of acid phosphatase production and rock phosphate solubilization by Bacillus haynesii strain ACP1. Sci Rep 12:8926. https://doi.org/10.1038/s41598-022-11448-6
43. Bakki M, Banane B, Marhane O, Esmaeel Q, Hatimi A, Ait Barka E, Azim K, Bouizgarne B (2024) Phosphate solubilizing *Pseudomonas* and *Bacillus* combined with rock phosphates promoting tomato growth and reducing bacterial canker disease. Front Microbiol 15. https://doi.org/10.3389/fmicb.2024.1289466
44. Wang J, Bai J, Xu J, Liang B (2009) Bioleaching of metals from printed wire boards by *Acidithiobacillus ferrooxidans* and *Acidithiobacillus thiooxidans* and their mixture. J Hazard Mater 172(2–3):1100–1105. https://doi.org/10.1016/j.jhazmat.2009.07.102
45. Wyciszkiewicz M, Saeid A, Malinowski P, Chojnacka K (2017) Valorization of phosphorus secondary raw materials by *Acidithiobacillus ferrooxidans*. Molecules 22(3):473. https://doi.org/10.3390/molecules22030473

Chapter 6
Productivity of Environmental Biotechnology Systems

Abstract Biotechnology products are very variable, and their diversity only increases with science and technological innovation development. This section deals with environmental biotechnology products, meaning biotechnologies for waste treatment and purification of environmental components. Since they aim to reduce the negative anthropogenic impact on the ecosystem. Their products also serve this purpose, including the biomass of microorganisms, which can be utilized in an environmentally safe way; biofuel as an alternative to traditional fuel, the use of carbon dioxide to reduce its emission into the atmosphere; the introduction of biofertilizers in agriculture to reduce its environmental impact. Most importantly, treated wastewater can be reused, waste can be disinfected and detoxified, and soil remediated. All these are ultimately products of biotechnology of ecological orientation.

6.1 Products of Environmental Biotechnology

The range of products obtained in environmental biotechnological processes is extremely wide. Regarding diversity and production volumes, the first place is occupied by products obtained in processes based on the activity of microorganisms. These products are divided into three main groups:

(1) *Microbial biomass*, as the target product (unicellular protein) or used as a biological agent in the processes of biomethanogenesis, bacterial leaching of metals, as well as microbiological plant protection agents, which are often a dried culture of microorganisms pathogenic to agricultural pests insects; biofertilizers; in the recycling of plant waste generated in various types of production (such as straw, press cakes, citrus waste, whey, molasses, animal manure, and domestic wastewater) [1].

(2) *Biochemical products*: primary metabolites (low-molecular compounds necessary for the growth of microorganisms as building blocks of macromolecules) and coenzymes which are used in food and feed and include amino acids, vitamins, acids, alcohols, and solvents; secondary metabolites (idiolites) are compounds that are not necessary for the growth of microorganisms and are not

Y. Chernysh et al., *Sustainable Biotechnology and Environmental Protection*, Green Energy and Technology, https://doi.org/10.1007/978-3-031-89436-7_6

associated with their growth, such as antibiotics, alkaloids, growth hormones, toxins, which are actively used in medicine and veterinary medicine [2].

(3) *Chemical products* formed as a result of transformations with the use of enzymatic activity of microorganisms—gaseous products of microbial metabolism (biogenic hydrogen sulfide, methane, ammonia)—biofuel production, technological processes of elemental sulfur production, technologies of heavy metal ions deposition in the form of sulfides [3].

(4) *Obtaining purified components of media from contaminants* (clarified wastewater, treated sludge from municipal wastewater, dewaxed oil, etc.) [4].

(5) *Precious metals* extracted from ores through microbiological leaching [5].

It is important to note that the duration of the main fermentation process depends on the specific target product to be obtained. Suppose the process is aimed at obtaining biomass. In that case, the purpose of fermentation is to obtain the maximum possible cell titer, and in the case of obtaining metabolites, their accumulation is carried out simultaneously, and the maxima of formation of the biological agent and target product are always shifted in time. Therefore, the duration of fermentation in the first case is always shorter than in the second. If the goal is to obtain biomass of the industrial strain in a batch process, the cultivation time does not exceed 24 h. During the production of primary metabolites, the biosynthesis time is 48–72 h, and for secondary metabolites—72 to 144 h.

The development of bio-based products is inherently a transdisciplinary and multifaceted task. Academia, industry, policymakers and the general public play their part in this process. At first, observing ethical and legal norms, the scientific community conducts systematic bioprospecting and screening of biological objects involved in biotechnology, finds out the structure of biologically active molecules and their mechanisms of action, and establishes product development protocols. Funding for bioproduct development depends on the needs and objectives of society (such as health, welfare or environmental protection) and the corresponding intellectual property protection (Fig. 6.1).

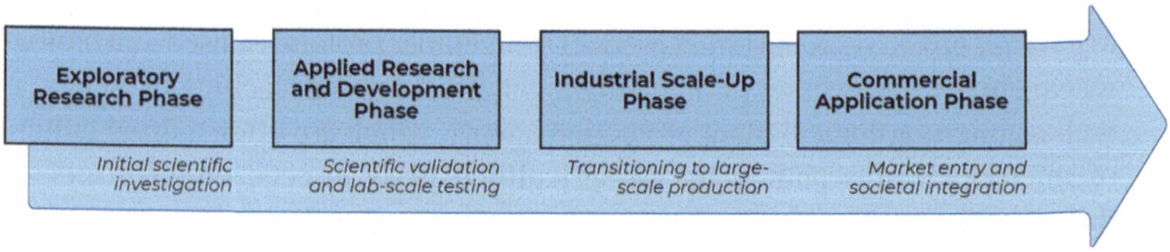

Fig. 6.1 Phases of biotechnological solution deployment

Table 6.1 The main characteristics of biogas containing CH_4 50–80%

Characteristic	Value
Density at standard conditions, kg/m^3	0.95–1.40
Lower heating value, MJ/m^3	18.0–27.5
Higher heating value, MJ/m^3	20.0–31.5
Ignition temperature, ^{0}C	650–750
Flammability limit (air content), %	6–12
Theoretical air volume required for combustion, m^3	4.8–7.6
Carbon dioxide content in dry combustion products, %	14.3–21.0
Normal flame propagation speed, cm/s	16–22
Concentration limits of inflammation, %	
Lower	6.5–10
Higher	17–31

6.1.1 Biogas

Biogas is obtained from organic raw materials in the process of anaerobic digestion as a result of the decomposition of complex organic substrates of various natures with the involvement of a mixed microbial association of different species microbial association, is a mixture of 65–80% methane and 20–35% carbon dioxide, as well as small amounts of hydrogen sulfide, nitrogen and hydrogen.

The use of biogas as a fuel allows the achievement of a positive effect. The products of energy processes associated with using traditional fuels account for 80–88% of all types of pollution in the biosphere. Table 6.1 presents the characteristics of biogas as a fuel.

The calorific value of biogas depends on the ratio of methane and carbon dioxide and is 5–7 kcal/m^3; 1 m^3 of biogas equals 4 kW/h of electricity, 0.6 l of kerosene, 1.5 kg of coal and 3.5 kg of firewood. Crude biogas are used in the home for heating homes and cooking and used as fuel in stationary plants that generate electricity. Compressed gas can be transported and used (after pre-cleaning) as fuel for internal combustion engines. Purified biogas are similar to natural gas. Biomethanogenesis solves not only the problem of energy recovery—but these processes are also fundamental in environmental terms, as they solve the problem of utilization and recycling of waste of various industries and technologies, agricultural and industrial, as well as domestic, including wastewater and municipal solid waste.

6.1.2 Biohumus

Environmentally friendly products obtained during the bioconversion of organic waste, under the action of various groups of microorganisms, both aerobic (composting, vermicomposting) and anaerobic (methanogenesis), are called **biohumus.**

The effect of different doses of biohumus on the microbiological activity of soils in the cultivation of spring wheat has been established. Biohumus has a slightly alkaline reaction (pH 7.4–7.8) with a total nitrogen content of 0.84–1.22%, phosphorus—from 0.69 to 0.99%, potassium—from 0.9 to 1.17% and mobile ammonium—from 232 to 347 mg/% of the substance. 1 g of biohumus formed during vermicomposting (a decomposition process with the help of earthworms) contains more than 378 million ammonifying bacteria and 251 thousand bacteria that decompose cellulose and mineralize organic matter. In biohumus experiments, the number of microorganisms in the soil was significantly higher than in the control. The largest number of ammonifying bacteria is achieved by introducing 10 t/ha of biohumus, while nitrates in the soil are less than ammonium. For example, from the moment spring wheat germinates until it is ready to be harvested, the number of nitrifying microorganisms (*azotobacter)* in the areas with biohumus increases in the soil. In this area, the activation of biological assimilation of atmospheric nitrogen by *azotobacter* is well noted. The content of bacteria that decompose cellulose in the soil increases until the tillering phase of spring wheat. Most of these bacteria are present in areas with biohumus [6].

The application of biohumus to the soil accelerates the mineralization of organophosphorus compounds as a result of the action of specific microorganisms. With the increase of norms from 10 to 30 t/ha the concentration of phosphorus-solubilizing bacteria increases. The phosphoric acid content in the soil depends on the number of microorganisms that decompose organic and mineral compounds of phosphorus [7].

The use of biohumus as a fertilizer affects the intensive growth of the microflora and accelerates the accumulation of mobile forms of nutrients needed to increase crop yields [8].

6.1.3 Biopreparations

Among the biopreparations used in agriculture, we can distinguish a large group of biofungicides, the basis of which are *Bacillus subtilis* bacteria, for example: phytosporin, bactophyte and others. Biopreparations are used to protect plants from phytopathogens throughout the growing season and to treat the fruit before storage.

Biopreparations used as biofungicides usually consist of selected natural strains of saprophytic bacteria that have a pronounced biological activity and are safe for all ecological niches (soil, plants, insects, animals, humans). When they enter the natural environment, such bacteria emit many antibiotic substances, enzymes, and other biologically active substances that slow down the development of phytopathogenic bacteria and fungi. Due to the high rate of reproduction and biological activity of bacteria that are part of biological products, they quickly assimilate the soil substrate, actively participate in the decomposition of organic compounds, ammonification and nitrification processes, and enhance the mobilization of phosphorus and potassium, enriching the soil with mobile nutrients. Biologically active substances released by them in a certain concentration stimulate the growth and development of plants, increasing their disease resistance.

Biopreparations based on bacilli, in addition to the protective function of plants against phytopathogens, have additional properties of probiotics for farm animals and poultry, which allows them to expand their practical application and increase their safety for animals and humans in obtaining feed and food from farm plants.

Bacterial treatment of plants, starting from soaking the seeds and throughout the growing season, leads to the accumulation of bacilli in the vegetative parts of the plant. When preparing feed, such a "bacterial additive" is a preservative for hay, as it prevents mould growth, helps maintain hay quality, and reduces the amount of heat released.

Biopreparations from bacilli can also be added as feed additives at the feed preparation stage. Feed additives—microbial probiotics, consisting of living microorganisms of the genus Bacillus, are added to feed and prevent their infection with phytopathogens and the accumulation of phytotoxins.

Thus, biofungicides from the genus Bacillus perform two functions simultaneously—the biofungicide itself (protects the plant from infection by phytopathogens of bacterial, viral or fungal nature) and probiotic—a microbial additive to feed for farm animals. Using Bacillus-based biopreparations biological products as biofungicides allows the production of plant-derived feed containing bacilli integrated into plant tissues. Using such feed for farm animals provides a high increase in animal biomass and more efficient use of feed, reducing the cost per unit of feed per unit weight.

The stimulation of growth and development of farm animals due to the presence of bacilli in feed occurs as a result of the multifactorial positive therapeutic and prophylactic effects of bacillus metabolites on the animals' organ systems.

Biologically active substances stimulate the growth and development of plants, increase their disease resistance, and are effective for all types of plants in the fight against: rust, gray and white mold rot, dry and brown fruit rot (phoma infections), brown spot (cladosporioides), septoria (white spot), coccomycosis, cytosporosis, moniliosis, phytophthora, root rot, grape mite, apple, pear and plum codling moths, leaf rollers, helminthosporiosis, black leg, powdery mildew, false powdery, fruit rot, scab, clasterosporiosis, oidium, bacterial, verticillium, bacterial, verticillium wilt and fusarium wilt, apple scale, spider mites of many species, and other leaf-eating pests.

Biological preparations are used throughout the growing season of plants, from pre-sowing treatment of seeds, seedlings, and saplings to final harvest.

For example, the biopreparation Guapsin is an aqueous suspension and is a universal biological preparation created on bacterial strains of *Pseudomonas Aureofaciens* (B-111) and strains B-306, which modify insoluble soil compounds in initial doses of macronutrients N, P, K. Guapsin is also an insectofungicide used to control pests and diseases of crops, which stimulates plant growth as a nitrogen-phosphorus fertilizer. Trichophyte biological product is an ecologically safe preparation based on a fungus of the genus *Trichoderma*. Fungi of this genus have long been known;

fungal strains, when multiplying, produce antibiotics that eliminate plant disease pathogens. Additionally, they destroy pathogenic fungi by using foreign mycelium as a food medium. Applying trichophytes improves the physicochemical properties of soils, including their structural condition, and the intensive processing of straw, stubble, and other plant residues. Trichophyte supplies soils with organic matter and affects humus formation and manure.

Research on crops and various soil samples that have been treated with biological products:

(1) Plant loss is significantly reduced due to root rot and powdery mildew.
(2) Stimulating the growth and development of plants.
(3) A significant increase in yield due to enhanced productive branching.
(4) Increase in yield to 80–98% through seed treatment.
(5) Strengthening the crop's resistance to frost and drought.
(6) Increase in yield by 20–50%.
(7) Reduction of the maturation period to two weeks.
(8) Longer shelf life of grains, vegetables and fruits treated with biological products.

Conditions with capillary moisture and temperature ranging from + 14 to + 25 °C are necessary for the active metabolism of the bacterial complex. It is recommended to carry out treatment with solutions of biological products in the evening or during overcast weather without bright sunlight. Also, temperature conditions, lower or higher than the optimal parameters, cause a decrease in the activity of microorganisms and, hence, their efficiency.

6.2 Criteria for Evaluating the Effectiveness of Biotechnological Processes

In biotechnology, when choosing a method to obtain a specific target product, it is necessary to conduct a feasibility study of alternatives to obtain such products by traditional methods. Compared to known biotechnological processes, they should be more technological, economical, and environmentally friendly, or alternatives should be excluded altogether. Evaluation of alternative options only at the cost of the product—one-sided. Evaluation of the effectiveness of biotechnology, in addition to the quality of the obtained products, can be a comparison of experimental and theoretical outputs of the product, calculated by the material and energy balance of the process. In large-scale biotechnological processes, the cost and consumption of raw materials are usually decisive factors, making material and energy assessments highly significant in this case. On the other hand, when using processes based on highly efficient recombinant biological agent strains, the leading share of the costs is not attributed to raw materials but rather to the creation of the biological agent and its maintenance, as well as the development of special conditions for its cultivation, i.e., in this case, the cost-efficiency of raw materials and energy resources is of secondary importance.

The biological agent plays a key role in any biotechnological process, including its nature and physiological and technological properties. The growth of any biological object requires viable seed material, energy sources and carbon, nutrients for biomass synthesis, absence of growth inhibitors, and appropriate physicochemical conditions of fermentation (pH, temperature, aeration, etc.).

The productivity of the process is characterized by the amount of product obtained per unit volume of the bioreactor per unit of time. **The productivity of the process** depends on many factors: the activity of the biological object, the value of the yield of the product from the consumed substrate, and the amount of active biomass in the fermenter:

$$P = q_s \cdot Y_{\frac{P}{S}} \cdot X, \tag{6.1}$$

where q_s is the rate of consumption of the substrate (metabolic coefficient), g/g h; $Y_p/_s$—product yield (economic coefficient), g/g; X—concentration of biomass, g/l; P—product; S—substrate concentration, g/l.

The fermentation process can be evaluated using various indicators and the appropriate equation.

Biomass productivity (Q_x, g/l h) is calculated by:

(a) for the batch process

$$Q_x = \frac{X_1 - X_0}{t_1 - t_0}; \tag{6.2}$$

(b) for a continuous process

$$Q_x = D \cdot X, \tag{6.3}$$

where X_0—biomass concentration (g/l) for a period of time t_0 (h); X_1—biomass concentration (g/l) for a period of time t_1 (h); D—the dilution factor or flow rate (1/h) equal to the specific growth rate (μ) for a continuous process; D can be disrupted if the flow rate of the medium or the concentration of the substrate in it is altered.

One of the main indicators characterizing the suitability of fermentation conditions is the growth rate of the biological agent. The growth rate (increase in biomass) of organisms with binary separation in a well-mixed medium in batch culture will be proportional to the concentration of microbial biomass and the time interval.

Specific growth rate (1/h):

$$\mu = \frac{X_1 - X_0}{X_1 \cdot (t_1 - t_0)}, \tag{6.4}$$

The formula can calculate the biomass concentration:

$$X_1 = X_0 \cdot e^{\mu(t_1 - t_0)}, \tag{6.5}$$

for the log phase of reproduction e = 2718.

The specific rate of substrate consumption (q_s, g/g) can be determined as follows:

$$q_s = \frac{S_0 - S_1}{X_1(t_1 - t_0)}. \tag{6.6}$$

The specific rate of target product formation (q_p, g/g h):

$$q_p = \frac{P_0 - P_1}{X_1(t_1 - t_0)}. \tag{6.7}$$

The value of productivity can be influenced by changing its various components, but in each case, it must be considered separately. Thus, with an increasing value of *X*, there may be restrictions on the mass transfer characteristics of the apparatus and limiting states; to influence the value of the metabolic coefficient of culture is possible only with a deep knowledge of the relationship between physiological and biochemical characteristics of the biological agent and environmental conditions.

The biomass yield from the substrate ($Y_{x/s}$), or the economic coefficient, can be calculated by the formula:

$$Y_{x/s} = \frac{\mu}{q_s} = \frac{X_1 - X_2}{S_0 - S_1}, \tag{6.8}$$

where S_0 and S_1 are the initial and final concentrations of the substrate.

Product yield ($Y_{p/s}$) (economic coefficient) is defined as the amount of product obtained from a given amount of substrate:

$$Y_{p/s} = \frac{q_p}{q_s} = \frac{P_1 - P_0}{S_0 - S_1} \tag{6.9}$$

This ratio expresses the substrate's efficiency in obtaining the target product. It is a very important characteristic because it is directly related to productivity and can directly affect the cost of the final product. The economic coefficient has a clear physical value, which characterizes the level of energy transfer from the substrate to the product.

This value is needed to calculate and predict the process as a whole. It is used as a parameter to monitor and control the progress of various processes and compare their efficiency.

The final concentration of the product should be planned considering the process's duration and the amount of product yield. Achieving the final high concentration of the product is justified if the selection and its concentration are time-consuming and expensive.

Specific energy consumption varies significantly depending on the direction and scheme of the fermentation process and the preparation conditions of raw materials at the pre-fermentation stage and post-fermentation procedures. Specific energy consumption also very significantly depends on the type of fermentation equipment. Unproductive substrate consumption (h) is the energy consumption of the substrate, which is not reflected in the growth of the product. In general, they are expressed through of economic coefficient:

$$h = Y_{experimental} / Y_{theoretical} < 1 \tag{6.10}$$

Unproductive costs significantly affect the efficiency and economics of the biotechnological process, so identifying the causes and locations of these additional costs of energy substrate is very important. Unproductive costs of the substrate may be associated with errors in reading genetic information in the process of rapid growth of the biological agent and the cost of maintenance in disconnected growth by reducing the efficiency of energy production in the electron transfer chain through the separation of oxidation and phosphorylation, inactivation of junctions, the emergence of alternative, less efficient branches with energy dissipation, as well as due to increased energy costs to maintain life without reproduction (transport of substrates and monomers in the cell, resynthesis of molecules, protective reactions and repair processes).

The initial evaluation of the efficiency of biotechnological processes according to the listed parameters is carried out at the laboratory development and the testing of the process stages. Then specified during scaling at the experimental and research-industrial stages. In comparative terms, anaerobic biotechnological processes are more economically advantageous when the costs associated with aeration, the operation of stirrers and bubblers, and others are sharply reduced. Consequently, the design of devices is simplified and becomes cheaper.

References

1. Srivastava RK, Sarangi P, Singh AK, Sahoo UK (2023) Microbial biomass production from biological processes. In: An insights to industrial microbiology. Daya Publishing House (A Division of Astral International Pvt. Ltd.), New Delhi
2. Fouillaud M, Dufossé L (2022) Microbial secondary metabolism and biotechnology. Microorganisms 10(1):123. https://doi.org/10.3390/microorganisms10010123
3. Wilke D (1999) Chemicals from biotechnology: molecular plant genetics will challenge the chemical and the fermentation industry. Appl Microbiol Biotechnol 52(2):135–145. https://doi.org/10.1007/s002530051500
4. Amapex (2020) Innovations in wastewater treatment: biotechnology and electrochemistry. https://amapex.net/wastewater-treatment-biotechnology/?lang=en
5. Banerjee I, Burrell B, Reed C, West AC, Banta S (2017) Metals and minerals as a biotechnology feedstock: engineering biomining microbiology for bioenergy applications. Curr Opin Biotechnol 45:144–155. https://doi.org/10.1016/j.copbio.2017.03.009

6. Khujamshukurov NA, Farhan M, Eshkobilov SA, Kuchkarova DX, Ahmad T, Fatima Z, Khan S, Akhtar J (2024). Impact of biohumus on soil agrochemical characteristics, fertility, and plant performance under greenhouse conditions. In: Talib F, Musa M (eds) Smart systems: methodological approaches and applications. CRC Press, 40 pp. https://doi.org/10.1003/495314
7. Charshanbiyev U, Xudyoberganov N, Odiniyev U, Yuldoshev O, Allanov A, Rakhmatullaev Y (2024) The role of biogumus in potato growing in the conditions of Uzbekistan. E3S Web Conf 563:03025. https://doi.org/10.1051/e3sconf/202456303025
8. Vinyukov O, Chuhrii H, Gyrka A, Vyskub R, Bondareva O (2022) Ways to improve the adaptability of winter wheat in the eastern part of the Northern Steppe of Ukraine. Univ J Agric Res 10(3):228–239. https://doi.org/10.13189/ujar.2022.100305

Chapter 7
Ecological Safety of Biotechnological Productions

Abstract The most important area of environmental safety is microbiological safety. In this direction, it is important to develop special engineering and technical support for biotechnological industries and the possibility of disposing of their waste. Wastes of biotechnological productions belong, as a rule, to the type decomposing in natural conditions under the influence of various factors (biological—mineralization with the participation of microorganisms; chemical—oxidation, physicochemical due to complex influence, for example, radiant energy and chemicals). Environmental safety includes the protection of human health from adverse factors of the natural and man-made environment, preservation of the natural environment and its restoration under adverse conditions, and development of new technologies with maximum economic efficiency and minimum impact on the environment.

7.1 Engineering and Technological Security of Biotechnological Industries

Biotechnological industries use a variety of raw materials and chemicals and produce a wide range of products in various forms. Several raw material components and some types of finished products, when released into the air of production facilities, can adversely affect the health of workers, create fire and explosive aerosols, and, in case of insufficient treatment of technological emissions into the environment—pollute the landscape at considerable distances from the enterprise. Both chemical and biological factors can determine the harmfulness of biotechnological production. Therefore, when designing biotechnology enterprises, special attention is paid to safety, industrial sanitation and environmental protection.

Industry standards have been developed for safety. In addition, all enterprises must compile the documentation in force at the plant on safety—general factory and general production instructions for jobs at all stages and types of work specified in the production regulations.

Regulations should contain complete data on waste, technological emissions, and effluents, as well as specific recommendations for their treatment, disposal,

Y. Chernysh et al., *Sustainable Biotechnology and Environmental Protection*, Green Energy and Technology, https://doi.org/10.1007/978-3-031-89436-7_7

or destruction. Of course, the same applies to all types of primary and secondary commodity products produced and sold by a biotechnology enterprise.

The most reliable way to ensure the biosafety of biotechnological industries is to organize production in accordance with the rules of asepsis.

Asepsis is a set of engineering and technological measures aimed at:

- Prevention of foreign microflora getting into the technological process, which ensures the efficiency of the technology and obtaining the product of the required quality.
- Prevention of the release of cultivated biological objects with air emissions and technogenic flows into the environment.

Technological processes according to the degree of asepsis are distinguished:

- *Aseptic*, which completely excludes the ingress of foreign microorganisms into the process and the release of the biological agent into the environment.
- *Conditionally aseptic*, which allows the presence of foreign microflora in the technological process and the entry of cells of the biological object in regulated quantities into the environment.
- *Non-aseptic*, all technologies that use pathogenic strains of microorganisms, viruses, and cell cultures of plants and animals, as well as the processes of obtaining products for medical purposes, are carried out under asepsis.

Special equipment and technology provide aseptic production conditions. In such processes, the penetration of even one foreign microorganism into the reaction volume of the fermenter means a violation of the technological regime. This can lead to a substandard product and the release of biological agent cells into the environment, leading to personal disease. Characteristics of sterilization methods are given in Table 7.1.

Conditional aseptic processes are carried out by:

- During solid-phase cultivation of a biological object to obtain biomass used for feed purposes, obtaining a biopreparations for bioremediation of polluted environments, and obtaining antibiotics.
- During submerged cultivation of a biological object on antiseptics substrates, such as methanol, acetic or lactic acid.
- During the cultivation of a biological object in conditions that are not optimal for developing infectious microflora (pH 4–5; pH 8–11; too low or too high temperature, etc.).

In technologies of conditionally aseptic processes, partial sterilization of incoming streams at is performed before the cultivation stage. This is usually the thermal sterilization of the original nutrient medium and compliance with all measures aimed at obtaining a pure culture inoculum and creating cultivation conditions conducive to the growth of the industrial strain. Therefore, the air supplied in these processes is not sterilized.

At anaerobic disinfection of liquid and solid organic wastes of various origins (municipal, industrial drains, solid household wastes, wastes of the agricultural

Table 7.1 Methods of sterilization

Method	Procedure	Application	Precaution	Limitation
Dry heat	Dry heat direct exposure at + 190 °C (80 min) or + 160 °C (130–160 min)	Glass laboratory equipment and products made from materials	High temperatures can damage thin-rolled or fine wire products	High temperature exposure can lead to unwanted changes in material properties
Autoclaving (high-temperature steam under pressure)	Three active components: temperature, steam and pressure + 121 °C (15 min) or + 126 °C (10 min)	Glass, fabrics, liquids under conditions where the components are resistant to high temperatures + 121 °C	It is not recommended for sterilization of most ordinary plastics	Necessity ensuring free air release from products before starting sterilization; usually used for small products
Gas (ethylene oxide)	Used individually and in a mixture with freon or carbon dioxide + 55–60 °C (2–3 h) or + 27–33 °C (5 h 30 min)	Almost any material, with some exceptions	Post-sterilization ventilation is required to remove gas residues that may be toxic	Ethylene oxide is a toxic and explosive gas; sterilization procedure is not economical
Gamma radiation (radioactive source—Cobalt)	Radiation emitted by the corresponding source (cobalt)	Wide use for sterilizing single-use products	Properties of some materials may change undesirably under gamma radiation exposure	Unwanted impact on properties of materials have a cumulative effect, and re-sterilization after it not allowed
Beta radiation (accelerator particles)	High-energy electron beam	–	–	–
Chemical (formaldehyde)	Substance in the form of vapour	Minimally used in industry	Toxic	Requires special equipment

sector), there is no need to use sterilization methods. During wastewater treatment in aerobic treatment systems, when treated effluents are discharged into water bodies, additional disinfection is necessary.

7.1.1 Systems for Purification of Gas and Air Emissions from Biotechnological Productions

Gaseous wastes in the processes of biological technology are few in range. The biochemical nature of enzyme-catalyzed reactions determines this. Usually, the energy source for biological agents is carbohydrates. Under aerobic and anaerobic conditions, carbon dioxide is formed from them. Thus, during fermentation (glycolytic process) from glucose, 2 mol of CO_2 are formed:

$$C_6H_{12}O_6 \rightarrow 2C_2H_5OH + 2CO_2C_6H_{12}O_6 \rightarrow 2C_2H_5OH + 2CO_2 \quad (7.1)$$

During the aerobic respiration (oxidative process)—6 mol of CO_2:

$$C_6H_{12}O_6 \rightarrow^{+6O_2} 6CO_2 + 6H_2O \quad (7.2)$$

During the growing methane oxidizing bacteria with CH_4 also formed CO_2:

$$CH_4 + 2O_2 \rightarrow CO_2 + 2H_2OCH_4 + 2O_2 \rightarrow CO_2 + 2H_2O \quad (7.3)$$

The released carbon dioxide is captured and utilized in the food industry as a refrigerant.

Gaseous waste from biotechnological industries based on the use of aerobic microorganisms is "exhaust air". It should not enter the atmosphere without cleaning and neutralization. Exhaust air is often a highly dispersed aerosol in which the dispersed phase is liquid droplets and/or microorganisms. The incidence rate of aerosol particles depends on their size and displacement of air flows. For some of them, the rates of falls are given in Table 7.2.

Table 7.2 shows that as the particle size decreases, their sedimentation significantly decreases, while Brownian displacement increases. That is why air currents can easily transport highly dispersed aerosols over long distances. Therefore, their adverse effects on sensitive groups of people inhaling microbial aerosols cannot be ruled out.

Table 7.2 Sedimentation rates of particles (density, g/cm) from highly dispersed aerosols (air viscosity—1.82×10^{-4} poise, t-293 K)

Particle radius, μm	Reynolds criterion	Sedimentation velocity, m/s	Brownian motion, m/s
3	0.366	0.11×10^{-2}	2.84×10^{-4}
1	1.43×10^{-2}	1.3×10^{-4}	5.07×10^{-6}
0.3	4.62×10^{-2}	1.39×10^{-5}	1.0×10^{-5}
0.1	2.45×10^{-5}	2.23×10^{-6}	2.1×10^{-5}
0.03	1.37×10^{-6}	4.16×10^{-7}	5.5×10^{-5}
0.01	1.26×10^{-7}	1.14×10^{-7}	1.06×10^{-4}

Exhaust air containing microorganisms (e.g. pathogenic exotoxin producers) must be heat treated and then undergo to filtration purification.

During the implementation of biotechnological processes, it is necessary to carry out constant (continuous) control over the quality of exhaust air released into the atmosphere.

Many biotechnological processes occur with the release of heat, which can be a by-product of biochemical reactions. Heat is often lost and not reused, except in a few productions where it is recycled.

The technological emissions of different biotechnology companies have very different physical and chemical properties and contents in terms of "biological factors". Based on this, when creating technical systems for the localization of emissions, there is a need to consider the following:

- Physicochemical properties of flows (temperature, humidity, pH, specific gravity, etc.).
- Quantitative and qualitative characteristics of biologically active particles (fractional dispersion composition, stability, etc.).

Biotechnological industries' main gas-air emissions (GAE) are aerosols, unstable dispersed systems consisting of small solid particles (cells, their aggregates) in the liquid phase, suspended in a gaseous environment.

All organized sources of GAE from fermenters, separators, flotators, centrifuges, dryers, packing sections and other technological equipment are equipped with cleaning systems inseparably connected with technological features of separate stages of the process.

The most common, proven methods of purification of GAE from biologically active particles are filtration, thermal treatment and "wet" purification systems.

According to its composition, the bulk of GPV can be divided into 2 groups:

- GAE containing living cells of microorganisms, drops of culture fluid with metabolic products and foam.
- GAE containing dry cells of microorganisms (protein dust) or small particles of the target product.

Purification of GAE of the first group provides a special separator for separating droplets and foam, followed by purification from the cells of microorganisms in the Venturi scrubber by 99.99%.

Purification of GAE of the second group, where the main component is protein dust, which reaches 200–300 mg/m^3, involves the technological scheme of a two-stage Venturi scrubber to ensure effective purification to the TLV value.

During the purification of gaseous emissions, the main task is the destruction of the aerosol. The time of destruction of aerosols is determined by the rate of sedimentation (deposition) of dispersed particles or the rate of their coagulation (increase due to the union of particles).

Most aerosol control methods are based on intensifying liquid particle fusion processes. To do this, porous ceramic or metal elements are often used.

In the first stage of purification of air emissions from fermenters, filters filled with metal shavings (mist eliminators) or Raschig ceramic rings are used at the beginning of the discharge pipe, where a partial increase of aerosol droplets occurs, followed by their return to the fermentation stage.

At the stages of fermentation and separation, the main air cleaning systems for the gases emitted by microorganism cells are Venturi scrubbers with a droplet separator or a two-stage system consisting of a Venturi scrubber and a mesh-type dust collector.

The stainless steel mesh of the knitted weave of the fog trap is placed perpendicular to the flow of the gas-air stream. Small droplets of aerosol containing cells of the industrial strain, colliding with the mesh, increase and flow into the lower part of the apparatus. The stability and efficiency of installation with a fog catcher are reached by means of continuous regeneration of a mesh of a fog catcher without interrupting of the main technological process.

During adherence to the technological regime, particularly the quality and quantity of irrigation water and the pressure drop across the gas phase, the "wet" cleaning system effectively reduces the concentration of microbial-containing particles in the technological gases that exit from fermenters (during cultivation of yeast) by an average of 4–5 orders of magnitude, ensuring a level below 102 cells/m^3. This ensures that the level of microbial contamination in the air is below the threshold limit value of microbial cells in the air of the work zone. The area of cell distribution is generally limited by a distance of 150–200 m from the emission source. The efficiency of such cleaning system is 99.9%.

In the biotechnology industry, a spray dryer is widely used. Spray dryers are vertical cylindrical devices with conical bottoms. The initial suspension or solution is dispersed into small droplets with a diameter of 10–100 μm using a disperse device. This leads to the development of an interfacial surface and the intensity of water vaporization in the stream of heated drying agents (100–150 °C).

The residence time of the material in the dryer is 15–20 s. The material is heated to a temperature of 40–60 °C (not higher than a wet bulb temperature). On the bottom of the apparatus, 70–80% of the dried particles settle and are removed by screw conveyor or pneumatic transport for packaging. The rest of the smaller particles are removed from the apparatus by the drying agent and captured by the gas cleaning system (GCS).

To obtain products where it is necessary to preserve the viability of microorganisms and high biological activity after drying (bread yeast, amino acids, antibiotics, enzymes, etc.), more gentle conditions are used for drying at temperatures of 40–50 °C. Since, under these conditions, the intensity of moisture evaporation decreases, the drying process is often carried out under a vacuum.

To ensure the sterility of the dried products, the drying agent (usually air) must be cleaned, for example, by sterilization filtration, before feeding to the drying chamber.

For obtaining products that do not require preserving the viability of microorganisms or providing their high biological activity, more intense and harsh drying regimes are used.

Various technological solutions are used to clean the gas-air streams of drying units, depending on the scheme of drying processes. A closed loop circulation of heat transfer carrier and air by pneumatic transport is also used.

Cleaning of gas emissions after spray dryers is also carried out in two stages: the first—is "dry" cleaning in cyclones, and the second is in Venturi scrubbers. The overall efficiency of such system reaches 99.5% on average.

To prevent the finished product from entering the air, several technological schemes for the drying process have been developed. In particular, a drying unit has been designed that includes a condenser for removing evaporated water. This allows the entire circulation loop to be closed in the scheme: air heater—drying chamber—cyclone group—"wet" cleaning apparatus—condenser—air heater (Fig. 7.1).

In this case, only part of the exhaust fuel gases from the air heater, which have no contact with dust particles of the dried product, is emitted into the atmosphere.

Such a system of cleaning allows to completely exclude the ingress of dry particles of the biological object in atmospheric air.

A desulphurisation system can be used to purify biogas from sulfur-containing impurities, which includes a scrubber and a biofilter with an immobilized accumulative or pure culture of thiobacilli (Fig. 7.2).

In the tubular heat exchanger, a liquid fraction of anaerobically digested waste is pumped through the pipeline as a heat carrier fluid, which has been cleaned and preheated in the boiler. The cooled liquid fraction from the tubular heat exchanger

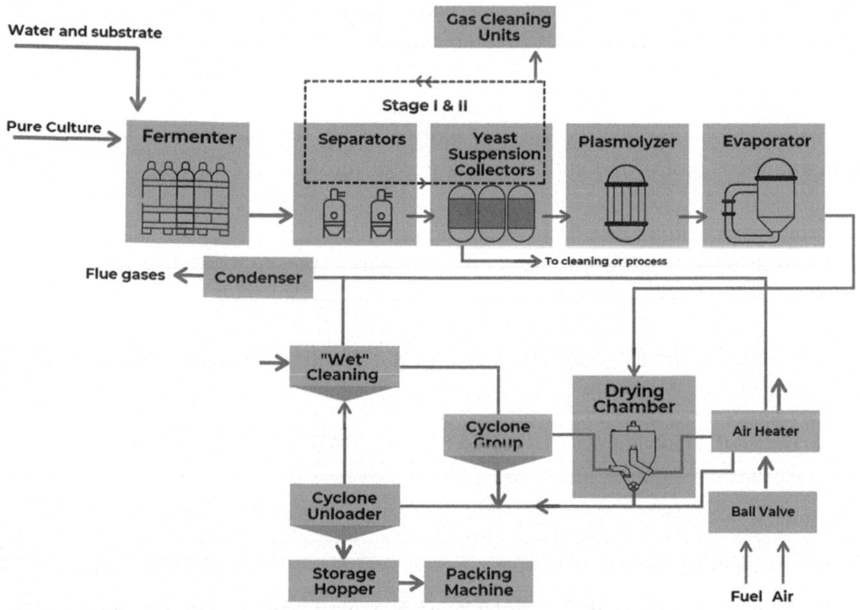

Fig. 7.1 Technological scheme using a closed circuit of exhaust air circulation

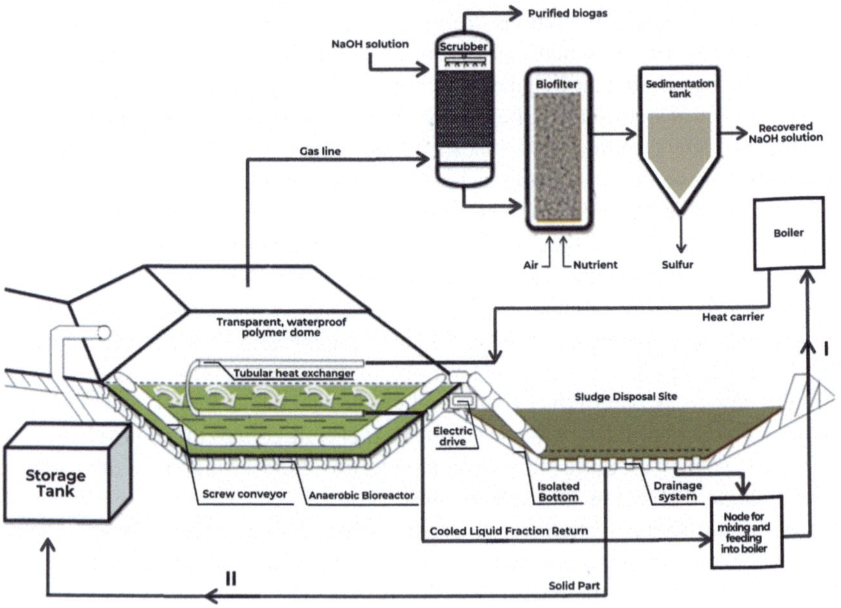

Fig. 7.2 Schematic biotechnological scheme of sewage sludge treatment [1]

returns to the node for mixing via the pipeline for the return of the cooled liquid fraction, where it is mixed with other liquid fractions and then sent to the boiler through line I. The separation of anaerobically digested waste into a solid and liquid fraction is carried out on the sludge disposal site, which has an isolated bottom and drainage system.

The solid fraction is an environmentally friendly organic-mineral fertilizer. At the same time, part of the solid fraction is sent through line II to the input of the bioreactor into the storage tank as a starter culture containing a formed sulfidogenic microbial association.

The following reactions in the scrubber occur at the supply of biogenic gas at a pressure higher than atmospheric (up to 75 bar) using caustic soda as a solvent:

$$H_2S + NaOH \rightarrow NaHS + H_2OH_2S + NaOH \rightarrow NaHS + H_2O \tag{7.4}$$

$$CO_2 + NaOH \rightarrow NaHCO_3CO_2 + NaOH \rightarrow NaHCO_3 \tag{7.5}$$

$$NaHCO_3 + NaOH \rightarrow Na_2CO_3 + H_2ONaHCO_3 + NaOH \rightarrow Na_2CO_3 + H_2O \tag{7.6}$$

$$H_2S + Na_2CO_3 \rightarrow NaHS + NaHCO_3H_2S + Na_2CO_3 \rightarrow NaHS + NaHCO_3 \tag{7.7}$$

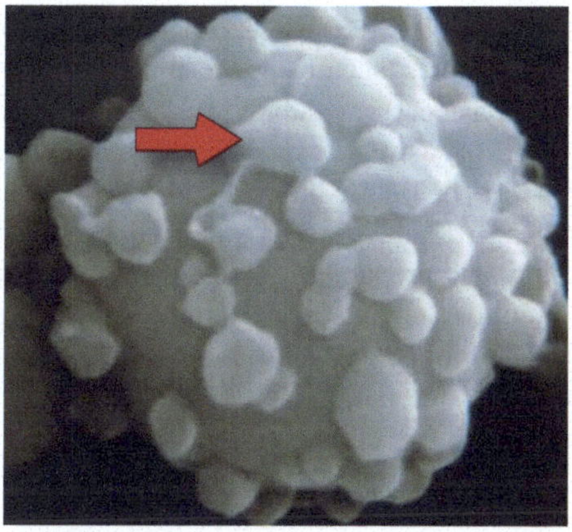

Fig. 7.3 *Theobacillus bacteria* with the formation of elemental sulfur on the surface (marked graphically by an arrow)

Dissolved hydrogen sulfide (HS–) from the scrubber is fed into the biofilter, where under atmospheric pressure and aerobic conditions, the following reactions of the biosulfurization process occur during bacterial metabolism:

$$NaHS + \frac{1}{2}O_2 \rightarrow S^0 + NaOH \tag{7.8}$$

$$2NaHS + 4O_2 \rightarrow 2Na_2SO_4 + 2H^+ \tag{7.9}$$

$$NaHCO_3 \rightarrow CO_2 + NaOH \tag{7.10}$$

It should be noted that the limited capabilities of different species of *Thiobacillus sp.* (for example, *Th. intermedius* and *Th. permetabolis*) to grow in autotrophic conditions are associated with the ability of these bacteria to use carbon dioxide to form various components of cells:

$$CO_2 + H_2O + \text{Enzyme} \rightarrow [CH_2O] + O_2 \tag{7.11}$$

The formation of sodium sulfate accounts for less than 5% of the total sulfide. The majority of hydrogen sulfide (95–96%) undergoes transformations, resulting in the formation of elemental sulfur (Fig. 7.3).

7.1.2 Wastewater Treatment Systems of Biotechnological Productions

Liquid waste in biotechnological industries is quite diverse in its composition. This is due to the incomplete use of biological objects of the components that are part of the nutrient media, the presence of substances (other than target metabolites) secreted by cells, the presence of solvents used, for example, for the extraction of final products, etc. Consider, for example, sulfite alkalis formed in the pulp and paper industry due to hydrolysis of wood and used to cultivate feed yeast. They contain, on average, 50–60% of lignin sulfonate, 7–8% of sugar sulfonic acids, about 18% of various sugars, 10% of sulfur dioxide, and 8% of calcium salts. After significant removal of sulfite and preparation of alkali (dilution, application of some nutrients), it is used for cultivating an adapted strain of yeast organisms.

The formed cell mass is the target product, and the waste is the culture medium after yeast separation. Liquid waste from yeast plants, which produce yeast on molasses wort, contains organic and mineral substances (mg/l on average): ethanol—0.45; carbohydrates (including fermentable ones)—1.0; total nitrogen—0.8; inorganic nitrogen—0.13; ash content—5,4. BOD of such waste is about 20,000 parts of O_2 per 1 million, i.e. about as much as BOD for sewage water. Waste generated from 1000 tons of molasses corresponds to the communal sewage water of the city, with a population of about 0.5 million inhabitants. Such liquid waste undergoes microbiological treatment (anaerobic or aerobic).

Industrial wastewater is heterogeneous. Thus, some can be considered relatively clean because they are almost similar to the natural water used in production (condensate, water from heat exchangers). Other waters are contaminated with inorganic and organic impurities that get: (1) from raw materials (pollution during transportation, washing potatoes, beets) and (2) from the equipment (washing of the technological equipment). Clean water can be reused in technological processes or directed to clean reservoirs; contaminated water is freed from mechanical impurities and then directed for treatment.

Also used in the production of antibiotics, besides water, are carbohydrates and carbohydrate products, oils, soy flour, corn extract, nitrates, ammonium salts, sulfur and phosphorus-containing compounds, possible precursors of antibiotics (e.g., amide phenylglycine as a precursor to penicillin; n-propyl alcohol as a precursor to erythromycin), inorganic acids and alkalis, organic extractants, etc.

A distinctive feature of biotechnological processes based on separating metabolites from culture fluids is the uneven ratio of the target product and the liquid (liquid medium) in which it is contained (often—1:100, 1:200, i.e. 1%, 0.5% and less). In such industries, the amount of liquid waste is significantlyhigher than solid waste. If the latter are contained in significant quantities in the wastewater, they are separated, compressed, and treated.

Industrial wastewater at state-owned enterprises contains 1–2% of organic matter and 98–99% of water. When using microbial treatment of such effluents in aeration tanks (for example, with the participation of saprophytic bacteria of the genus *Afthrobacter, Corynebacterium, Mycobacterium, Nocardia* and some others), valuable microbial fertilizers can be obtained.

All liquid effluents of production can be divided into primary and secondary production streams.

The wastewater of the primary production includes, the spent culture fluid formed during the separation of biomass from the liquid phase, which constitutes 80% of the total production effluent. This group also includes irrigation water flows of the gas cleaning system, Venturi scrubbers, wash water from the separation stage, water from washing equipment, and water from cooling systems.

Effluents of secondary production, after treatment, can be used for supply to recirculating cooling water systems in water supply production, as well as for compensating water losses during evaporation and droplet emission on condensers.

For processing contaminated by microorganisms industrial effluents, chemical, biochemical, and physical methods, as well as their combinations, are used.

Chemical methods are based on introducing into the effluent a suitable chemical reagent at an optimal concentration under optimal environmental conditions: temperature, mass transfer, pH duration of exposure, etc. It is established that chemical methods of inactivation of even vegetative forms of microorganisms require significant concentrations of chemical reagents during the whole process, adequate mixing with effluents, elevated temperatures and prolonged exposures.

After completion of the treatment of effluents with chemical reagents before discharging them into the sewer and sending them for biological treatment, the excess unreacted chemical agent must be neutralized to maintain the required activity of the microflora of biofilters or activated sludge.

All this makes chemical treatment unreliable and uneconomical, so it has not become common in microbiological industries.

Among the physical methods that can be used are ionizing radiation, ultraviolet radiation, ultrasound, reverse osmosis and ultrafiltration, and heat treatment.

Thermal steam decontamination of effluents can be carried out in systems operating on a cyclic or continuous-flow principle.

The following equation determines the number of devices (n) of a certain capacity:

$$n = \frac{Q \cdot \tau}{24 \cdot Y \cdot K_3} \tag{7.12}$$

where Q—the daily amount of runoff, m; Y—the total capacity of the unit for wastewater collection and heat treatment; K_c—the filling factor of the unit (for processes without foaming—0.7–0.85, with foaming—0.4–0.6); τ—duration of one cycle, including operations of inspection and preparation of the unit, wastewater intake, their heating to processing temperature, holding, cooling and discharge into the sewerage, h.

A typical steam treatment mode involves heating to 128 °C, holding for 30 min (sometimes up to 1 h) and cooling to 40 °C.

The disadvantage of cyclic devices is their relatively low performance.

The more widely used continuous flow method is designed for rapidly heating liquids in a continuous stream to the neutralization temperature, holding for seconds or minutes and rapid cooling.

During steady-state operation, the effluent from the collector is pumped into the recuperator, where it is preheated through heat exchange with hot wastewater discharged from the system after retention at the treatment temperature, and then directed into the steam heater. The effluents heated to the decontamination temperature pass through the holder, then through the recuperator and cooler, after which it is sent to biological treatment. The temperature at the outlet of the holder is blocked by valves so that if the temperature is lower than required for decontamination, the effluent on the recirculation circuit is returned to the effluent collector, not discharged.

The generally accepted wastewater treatment methods of industrial enterprises, including biotechnological ones, are biological, based on the ability of microorganisms to use wastewater pollution as a power source.

The greatest technical difficulty in the treatment of wastewater from enterprises producing biomass of microorganisms is their release from nutrients—nitrogen, phosphorus and anions of strong acids (sulfates and chlorides), which are part of the nutrient medium during the cultivation of the biological agent. According to experts, the cost of wastewater treatment can reach 30–40% of the cost of the enterprise.

Based on this, it is evident that increasing the efficiency of the purification of liquid streams of biotechnological industries is associated with improving the basic technological process. One such approach is the reuse of spent culture fluid at the stage of cultivation, which reduces the total consumption of water consumption in production, raw materials (mineral nutrients) and the flow entering the biological treatment stage. However, the volume of culture fluid returned is limited (no more than 70%) due to the accumulation in the culture medium of metabolic products.

The fundamental solution to this problem is to develop a process for obtaining biomass of microorganisms in a closed cycle of water use. Implementing such technology requires regulation of the composition of the nutrient medium, especially the regulation of the supply of nutrients to the fermenter.

Biological wastewater treatment is considered the second fermentation stage during the closed cycle of water use. Modes of supply of elements of mineral nutrients, sewage treatment, and return of the cleared liquid streams to a fermentation stage in the conditions under which a uniform technological system is developed.

The implementation of a closed system of water use at the plant for the production of feed yeast from n-paraffins allowed the complete elimination of the discharge of wastewater into the natural reservoir, reduced the consumption of fresh water for technological needs from 130–160 m^3 per 1 ton of product to 14 m^3/t.

To ensure environmentally friendly biomass production of microorganisms, it is necessary to:

- prepare a nutrient medium using salts that prevent sludge formation (e.g., liquid ammonium phosphate) to eliminate solid waste.
- exclude industrial wastewater discharge by returning spent culture fluid or biologically treated water into the production process.
- use drying and granulation units with a closed circuit of coolant circulation to eliminate the ingress of dust from the finished product into the atmosphere.
- exclude dusting during loading and unloading operations on transport and at the consumer due to granulated product release.

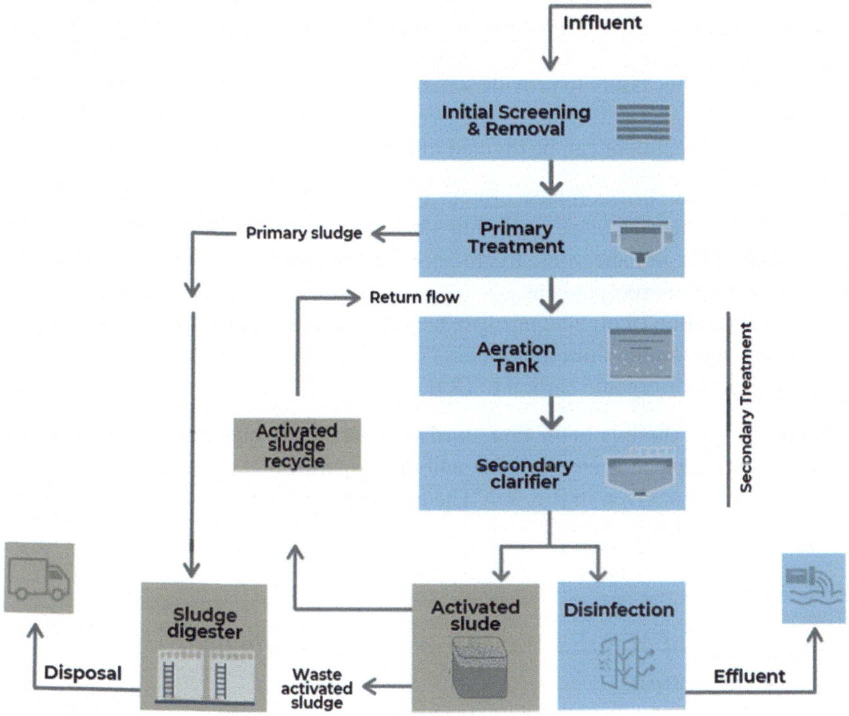

Fig. 7.4 Scheme of biological wastewater treatment

Depending on the wastewater quality, it is also possible to treat it appropriately (for example, to obtain return water, which is re-sold in the same biotechnological production). Figure 7.4 shows the scheme of biological wastewater treatment.

Dissolved organic substances can be removed using activated sludge in aeration tanks or aerobic treatment, on biological drop filters; nitrates are neutralized by denitrifier microbes (*Pseudomonas* spp., *Vas. licheniformis, Paracoccus denitrificans, Thiobacillus denitrificans*), phosphoric acid salts are coagulated and precipitated. The newly formed solid (thick) precipitates are concentrated, dehydrated (by filtration, centrifugation, settling on a sand layer), and then burned or used as fertilizer.

Thus, during the treatment of wastewater to the level of pure water, the following phases can be distinguished: separation of large particles that easily settle and oil films (rough treatment), separation of suspended particles and dissolved organic matter (moderate fine treatment) and, finally, separation of all other impurities (fine purification). At rough cleaning, particles of the size of 100 μm and more are separated, and at moderately thin—from 1 to 100 μm, and at thin—from 0.1 to 1 μm.

Deep purification of wastewater is consistently achieved by filtration through sand layers, chlorination, filtration through activated carbon, evaporation (liquid extraction, freeze crystallization, reverse osmosis), and ion exchange.

If precipitation (solids) is formed in this phase, they are added to other precipitates and treated as described above. In all cases of organization of biotechnological productions, it is necessary to provide separate systems of drains—technological and municipal.

Thus, waste treatment can be divided into 4 stages:

(1) The destruction of complex protein complexes (mainly—their conjugates) to simple soluble substances and their separation from insoluble substances.
(2) Liquefaction and anaerobic treatment of the insoluble residue with the help of appropriate microorganisms.
(3) Transformation of organic nitrogen to NH_4^+ (ammonification) with subsequent oxidation of ammonium to nitrates.
(4) Conversion of organic carbon to CO_2.

In these stages (mainly in the first, third and fourth), the biochemical nature of the processes taking place is emphasized, and the generalized technological scheme of processing waste is given in Fig. 7.4. The insoluble residue (stage two) can be used to produce methane.

7.1.3 Disposal of Solid Waste from Biotechnological Productions

Solid waste in biotechnological industries is a microbial mass separated from the culture filtrate, which enters the following stages of isolation of the target product: sludge; plant biomass after extraction of active substances (and in the case of suspension culture, which produces metabolites in the nutrient medium, the waste is cells); residues of chicken embryos in the cultivation of, for example, influenza virus; some tissue cultures of mammals; precipitation from sewage (silt). It is estimated that in municipal treatment plants, one person generates about 500 l of sludge per year with a moisture content of 5%. If the urban population in a country is 100 million people, then in 1 year, 47.5 million m^3 of such sludge will accumulate. Adding almost the same amount of industrial sludge, including solid waste from biotechnological production, it is requires significant efforts and resources to neutralize or dispose of them.

Long-established biotechnological productions in many countries include industrial methods of obtaining beer, yeast, wines, etc. Using the example of brewing, it can be noted that solid waste here is yeast cells (0.25–0.40 kg per 1 sip of beer), malt and hop dregs, and protein precipitate from separators. Residues of hops (hop dregs) and protein contain bitter substances, so they are not used as additives in animal feed diets. Therefore, such residues are either incinerated (which is unprofitable) or transferred for biological treatment. Under optimal environments and aeration, the biomass of filamentous fungi and yeast cells can be 2.5% in terms of dry weight, and about 50% of it is protein.

To some extent, the quality of solid waste dictates the choice of disinfection method. Thus, pathogenic microbes—that produce potent poisons (toxins)—must be completely neutralized, and the most effective way to do this is by burning. If the waste is biomass of streptomycete cells, it is sufficient to destroy them by heating with subsequent removal to farms where they can be added to animal feed (e.g., solidified waste in tetracycline antibiotic production containing proteins and vitamin B12), or mixed into soil as an organic fertilizer; sent to municipal wastewater treatment plants, and also to fermentation.

If, according to the technological scheme, solid and liquid wastes are delivered in the form of mixed runoff, then first carry out the rough separation of the former from the latter, then carry out the extrusion of moisture with subsequent transfer of compacted cell biomass for disposal in the above ways.

Similar approach can be taken to solid waste of plant or animal origin—toxic incinerated, non-toxic if possible sent for disposal.

When disposing of solid waste in microbiological production only by its destruction, it is necessary to consider the antigenic features of such microbial biomass (ability to cause the formation of antibodies in vivo). In any case, excluding the sensitizing (sensitive) effect on the microorganism is necessary. avoidance of possible allergic diseases [2].

In the aeration tanks of wastewater treatment plants, where the waste is disposed of, the limited factors are mainly the quality and area of the biological film, consisting of micro and macro flora and fauna. In this regard, it is necessary to be sure that the added solid waste, rich in organic matter, will not impair the performance of aeration tanks.

In anaerobic methane fermentation, virtually any organic matter (except lignin) can be a substrate transformed into methane and carbon dioxide. Methane is used as a fuel, and carbon dioxide is used in the food industry as "dry ice". Solid residue after methane fermentation (approximately 40% of the initial amount) is biohumus used as fertilizer in the cultivation of crops.

7.2 Assessment of the Sanitary-Microbiological State of the Environment of Biotechnological Productions

Biological aerosol, which accompanies biotechnological production, is the main "biological factor" source. Biological aerosol is an aerosol in which the dispersed phase consists of particles with inactivated or viable cells of microorganisms, products of their metabolism, spores of microorganisms, and plant dust.

The main potential danger when working with microorganisms is inhaling them through the air. When working with actinomycete spores, it was found that particles with a size of 0.02–0.2 μm can get into the deep parts of the lungs (alveolus).

Suspension microbial cultures easily form an aerosol. Drops with a diameter of 0.1 μm settle quickly and dry for a long time, 0.4 s, while their diameter decreases, and air currents transport them.

Aerosol particles with a diameter of < 10 μm are retained in the body by 100%; < 5 μm—by 20–30%; < 0.1 μm—by 60%.

During the sanitary and hygienic studies of the air environment, the number of cells of the biological agent culture, the total contamination with microorganisms and their group composition, the amount of protein, and the content of other chemicals are determined. The purpose of sanitary and hygienic studies is to:

- Identification of sources of air pollution.
- Determining the composition of organized emissions, general ventilation and unorganized production emissions.
- Determining the dependence of the degree of air pollution on the characteristics of the technological process, equipment, and effectiveness of collective protection.
- Develop and implement measures to improve working conditions and prevent environmental pollution.

The criterion for assessing the sanitary-microbiological condition of the air of the work zone is the maximum allowable concentration (TLV) or the approximate safe level of exposure used in the production of the strain.

The microbiological state of the air is assessed based on dynamic studies of the concentration of microorganisms.

The results of control of air pollution in industrial premises and organized emissions determine the observance of hygienic standards, TLV of microbial cells in the air of the work zone, and quantitative characteristics of the efficiency of equipment that retains and inactivates the microflora.

The main sanitary and hygienic characteristics of air in biotechnological productions are concentrations of production strain, foreign microflora, and species composition.

Control of foreign microflora is especially important for technologies based on unprotected fermentation processes, using raw materials that are well absorbed by many microorganisms [3].

Technological equipment is the source of air pollution by microorganisms in the work zone. For example, the study by Wang et al. [4] showed that during the production of microbiological plant protection products, such as beauvericin, the active ingredient of the fungus *Beauveria bassiana*, significant air pollution can be observed in the work zone due to the production strain. The study showed that even before work begins, the air can still be heavily polluted with spores of various fungi—up to 10^4 spores of *Beauveria bassiana*, 10^3 spores of Penicillium, and up to 10^2 spores of *Aspergillus niger* were detected in 1 m^3 of air. During the working day, the concentration of spores of the fungus *Beauveria bassiana* in the air of industrial premises increased 5–7 times [4].

During the production of the feed additive of bacitracin, the air pollution in the production room averages 10^3 microbial bodies in 1 m^3 of air. The number of staphylococci can be up to 15–16 microbial bodies; mold can also be detected in 80–85 microbial bodies, and the production strain can be up to 10^3 microbial bodies. The maximum microbial contamination (mainly production strain) can be observed in the fermentation and packaging department—up to 10^4 microbial bodies.

The method of determining microorganisms in the air of the work zone is based on the calculation of colonies that grew on agar nutrient medium during the deposition of microorganisms from the air by special devices that capture microbial aerosols (e.g., slit apparatus Krotov, impactors, etc.).

Sampling of air for bacteriological analysis is carried out by the method of inertial deposition, which provides air aspiration up to 20–40 l/min and sampling time—up to 30 min. The capture efficiency of the device is 50% for particles with a diameter of 3.6 μm, up to 95% for particles with a large diameter [5].

The sampling device is installed at the control point on a flat horizontal surface in the breathing zone at a height of 1.5 m. In addition to the air environment, in each control point, sanitary-microbiological studies of the surfaces of technological equipment, walls, and overalls can be conducted, and dust samples from the air can be collected simultaneously to determine the protein content [6].

References

1. Chernysh Y, Plyatsuk L, Dorda V (2014) Ecotechnology for hydrogen sulfide removal and production of elemental sulfur. Int J Energy Clean Environ 15(2–4):189–202
2. Ezeonu CS, Tagbo R, Anike EN, Oje OA, Onwurah IN (2012) Biotechnological tools for environmental sustainability: prospects and challenges for environments in Nigeria—a standard review. Biotechnol Res Int 2012:450802. https://doi.org/10.1155/2012/450802
3. Chawla H, Anand P, Garg K, Bhagat N, Varmani SG, Bansal T, McBain AJ, Marwah RG (2023) A comprehensive review of microbial contamination in the indoor environment: sources, sampling, health risks, and mitigation strategies. Front Public Health 11:1285393. https://doi.org/10.3389/fpubh.2023.1285393
4. Wang H, Peng H, Li W, Cheng P, Gong M (2021) The toxins of Beauveria bassiana and the strategies to improve their virulence to insects. Front Microbiol 12:705343. https://doi.org/10.3389/fmicb.2021.705343
5. Chiyaka TL, Nyawo GR, Naidoo CC et al (2024) PneumoniaCheck, a novel aerosol collection device, permits capture of airborne *Mycobacterium tuberculosis* and characterisation of the cough aeromicrobiome in people with tuberculosis. Ann Clin Microbiol Antimicrob 23(74). https://doi.org/10.1186/s12941-024-00735-x
6. Haig CW, Mackay WG, Walker JT, Williams C (2016) Bioaerosol sampling: sampling mechanisms, bioefficiency and field studies. J Hosp Infect 93(3):242–255. https://doi.org/10.1016/j.jhin.2016.03.017

Chapter 8
Phosphogypsum in Bioprocesses Application

Abstract The phosphogypsum is a by-product of phosphoric acid production for phosphate fertilizer. The problem of its accumulation in the environment is becoming increasingly important due to large-scale fertilizer production. This Chapter evaluates the potential of phosphogypsum as a mineral additive for the microorganism's growth under biotechnological processes. This considers phosphogypsum's component composition and impurities, which may vary depending on the production technology and the phosphorus feedstock (apatite and phosphorite).

8.1 Issues of Phosphogypsum Deposition and Its Application

Phosphogypsum (PG) is a largely underutilized by-product of phosphate fertilizer production, containing valuable components such as calcium sulphates and rare-earth elements, including silicon, iron, titanium, magnesium, aluminum, and manganese. However, it also harbors toxic substances, including heavy metals and other contaminants. The conventional disposal method for phosphogypsum involves open-air storage in large waste dumps, often located near industrial facilities, natural ecosystems, and residential areas. These dumps occupy vast areas and pose environmental risks. Additionally, the transportation and storage of phosphogypsum impose significant financial burdens, accounting for up to 10% of the total cost of phosphoric acid production. As industrial processes continue to generate phosphogypsum waste, these disposal sites are expanding at an alarming rate, threatening the stability of surrounding ecosystems. The global accumulation of phosphogypsum is estimated to increase by approximately 200 million tons annually, yet only 10–15% of this by-product is currently being utilized.

The issue of phosphogypsum (PG) utilization and storage is a significant environmental challenge for many countries, as it contributes to the pollution of water, soil, and air. In Ukraine alone, more than 55 million tons of phosphogypsum have accumulated in various locations, including Armyansk, Sumy, Rivne, and other cities, occupying vast areas that could otherwise be used for agriculture. The Sumy region

Y. Chernysh et al., *Sustainable Biotechnology and Environmental Protection*, Green Energy and Technology, https://doi.org/10.1007/978-3-031-89436-7_8

alone holds over 15 million tons of phosphogypsum waste, while anthropogenic phosphogypsum deposits in the Rivne region amount to 15.2 million tons. Additionally, 421,110 tons of phosphogypsum have accumulated in the Vinnytsia region. Due to its radioactive properties, phosphogypsum disposal is strictly regulated in some countries. The U.S. Environmental Protection Agency (EPA) prohibits most uses of phosphogypsum, citing concerns over its radium content. For instance, phosphogypsum from Central Florida contains an average of 26 pCi/g of radium, while samples from North Florida average 10 pCi/g. However, an exception is made for phosphogypsum with a radium concentration below 10 pCi/g, which can be used as an agricultural soil amendment, though its application remains restricted in other sectors.

China also has significant production of phosphogypsum, and the five largest provinces in terms of production are Hubei, Yunnan, Guizhou, Sichuan and Anhui; hence, they also have large storage sites. Poland also has significant phosphogypsum storage facilities. More than 5 million tonnes of apatite phosphogypsum has been accumulated in the Chemical Wizow Plant (Poland) waste dumps. These wastes contain rare earth elements, listed as "critical" raw materials in the EU.

Phosphogypsum dumps pose significant environmental risks, contributing to the contamination of groundwater, surface water, soil, and vegetation. Harmful substances can leach through protective barriers due to evaporation, runoff from precipitation, and airborne dispersion caused by wind erosion and dust formation. These dumps also exert a hydrodynamic impact, altering groundwater levels, which can lead to land pollution, land alienation, and disruptions to the natural landscape, particularly in residential areas. The large-scale disposal of phosphate fertilizer production waste also has severe consequences for marine ecosystems. Research by El Kateb et al. [1] highlights the disruption of the phosphorus cycle due to phosphogypsum discharge, particularly in the Gulf of Gabes (Tunisia), where high levels of authigenic phosphorus have been detected. Similarly, Gaudry et al. [2] found that phosphogypsum, primarily composed of calcium sulphate and other salts, releases heavy metals into seawater as it dissolves, impacting marine environments near discharge sites. This is especially concerning in the Atlantic Ocean, where the Moroccan phosphate industry releases substantial amounts of phosphogypsum waste near Safi and Jorf Lasfar. Additionally, research by El Zrelli et al. [3] examined the formation of foam layers resulting from phosphogypsum discharge into coastal waters. These floating layers can be passively transported over long distances, influencing pollutant distribution and further exacerbating marine contamination.

8.2 Current Trends in the Use of Phosphogypsum

Considering the quantities of formed and produced phosphogypsum, the problem is not only the removal, transportation and storage in waste dumps and sludge collectors but also its utilization with already developed methods and the development of new methods and more possibilities for its use.

Based on the number of publications (Fig. 8.1), the interest in phosphogypsum as a source of secondary raw materials has significantly increased over the past decade. Initially, phosphogypsum was mainly considered a component for construction, cement, road-building, and agricultural industries. However, the focus has shifted over the last 10–20 years, given the increasing anthropogenic pressure on the environment and the resulting shortage of natural sources of raw materials. Phosphogypsum, which has many useful elements, is therefore considered a source of calcium, phosphorus, rare-earth elements, trace elements, etc., as well as a mineral resource in technological environmental protection processes.

There are various directions and ways to directly use phosphogypsum and its processing in other products. These have proven technical feasibility and expediency for using phosphogypsum in the national economy instead of traditional raw materials [5].

According to the aforementioned properties of phosphogypsum, the field of research is currently (Fig. 8.2) shifting more towards environmental sciences, but engineering research is simultaneously being actively developed. Research in materials science is also being developed due to the component composition of phosphogypsum.

The scope of application of phosphogypsum is presented in Fig. 8.3.

The use of phosphogypsum in agriculture for soil fertilization and reclamation, particularly for saline soils, remains a promising approach. This is especially relevant in the context of changing climatic conditions and increasing aridification, which contribute to the expansion of anthropogenic (secondary) saline soils. Beyond agriculture, phosphogypsum is gaining attention in biotechnology and environmental protection. It serves not only as a substrate for microbial cultures but also as a nutrient-rich medium in biotechnological applications. These include off-gas cleaning, wastewater treatment, and the remediation of soils contaminated with petroleum products, drilling cuttings, and other pollutants. The integration of phosphogypsum in such technologies enhances its value as a multifunctional resource.

A visualization of research trends and interconnections related to phosphogypsum utilization across various fields was generated using the VOSviewer program, highlighting key clusters and areas of application (Fig. 8.4).

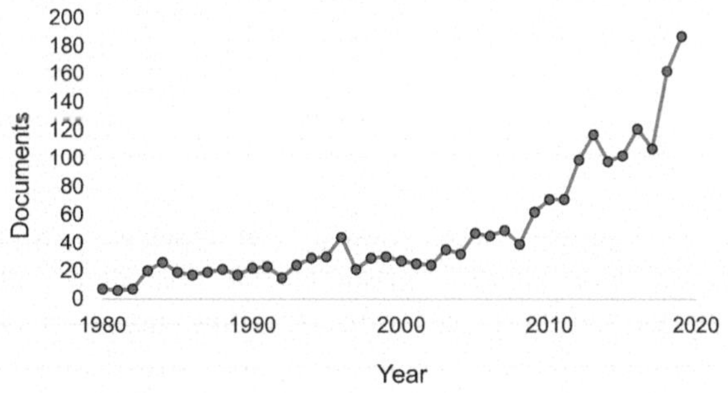

Fig. 8.1 Publishing trends (number of papers) in phosphogypsum research (using Scopus Database) [4]

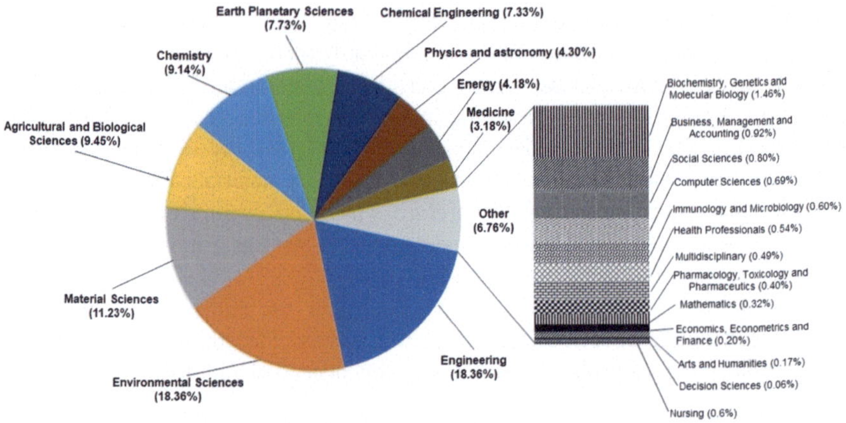

Fig. 8.2 Papers in the field of "phosphogypsum" classified by fields (Scopus Database) [4]

Fig. 8.3 The main directions of phosphogypsum recycling [4]

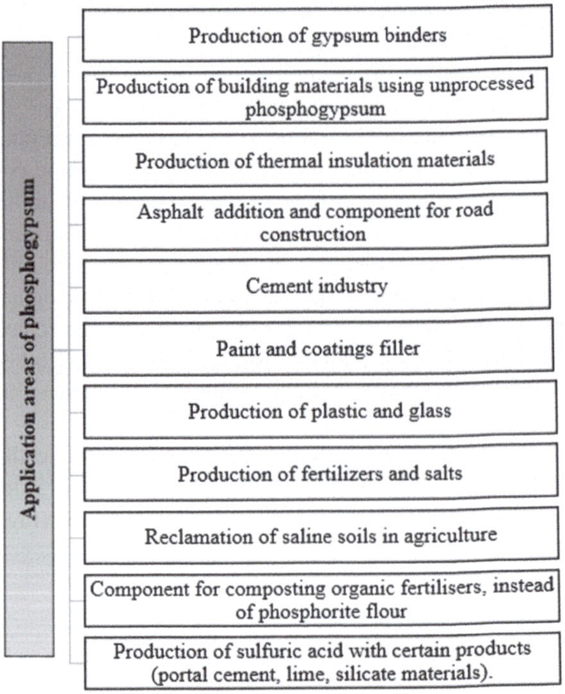

Five clusters were formed: **red cluster**—phosphogypsum recycling in the construction industry; **green cluster**—radiation pollution problem of phosphogypsum and phosphate fertilizers; **yellow cluster**—monitoring migration of phosphogypsum components in the ecosystem, with the mobile forms of heavy metals and

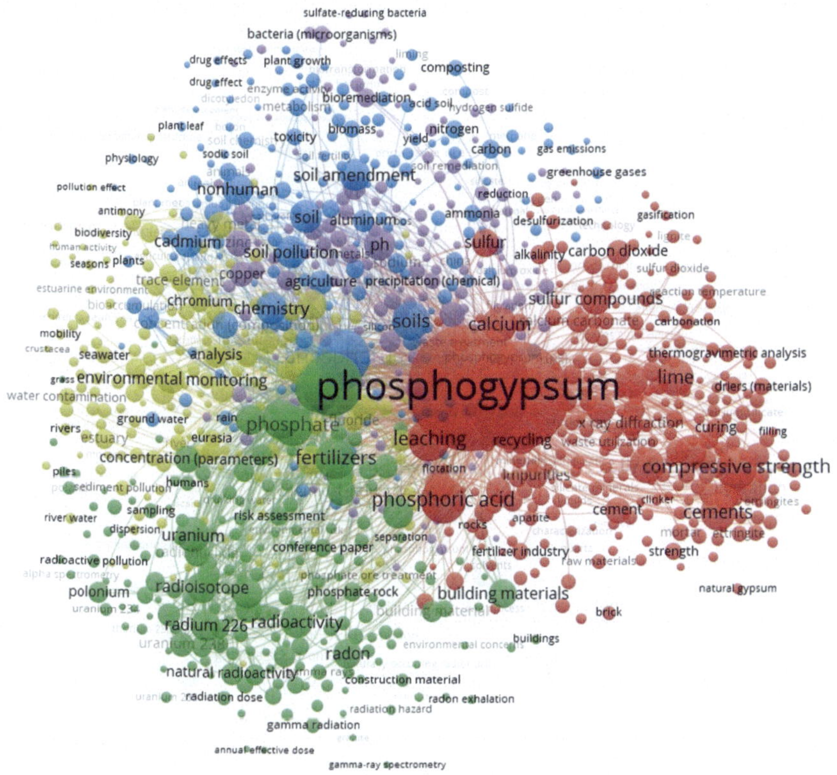

Fig. 8.4 General network visualization (constructed via VOSViewer v.1.6.15): 5 clusters, 94,846 links, 209,564 total link strength [4]

their inflow into aquifers from phosphogypsum dumps; **blue cluster**—use of phosphogypsum in agriculture as an ameliorant and a component of complex fertilizer; **purple cluster**—the impact of phosphogypsum on microorganisms, particularly in bioremediation processes.

The analysis of phosphogypsum component composition is important in assessing the environmental safety of phosphogypsum use (Fig. 8.5). Phosphogypsum of different origins and storage times has a variation in the composition of impurities, which significantly affects the scope of its possible application. The composition of phosphogypsum is determined by the initial raw material of phosphorite from which phosphate fertilizers are produced.

As can be seen from the highlighted fragment of visualization (Fig. 8.5), the main components that are important for the implementation of certain technological solutions of phosphogypsum conversion are calcium sulfates in a semi-gyrate or dihydrate form, undecomposed residues of phosphates and the potential presence of heavy metals and radionuclides.

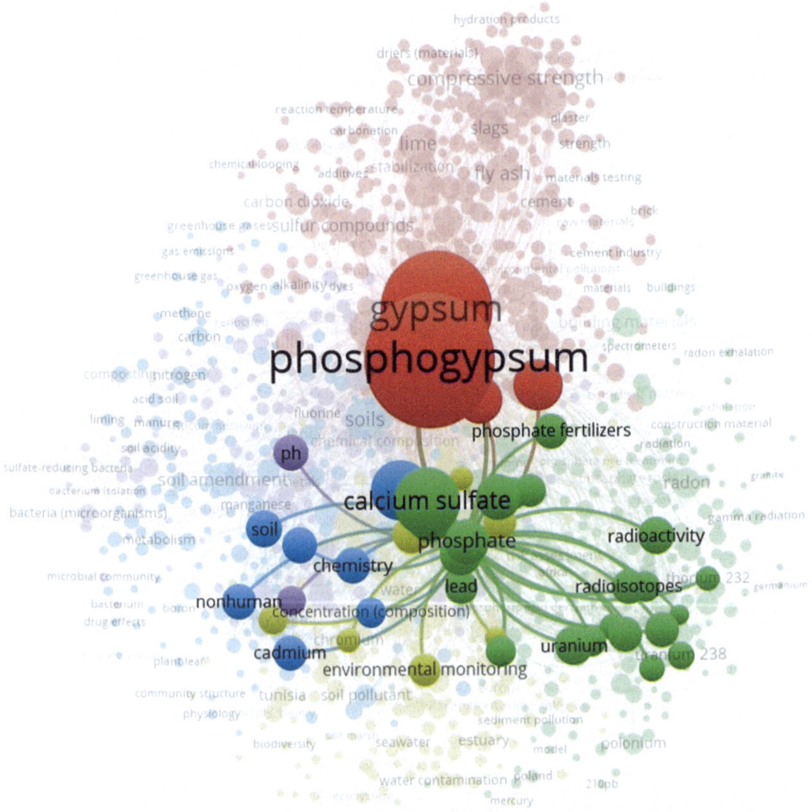

Fig. 8.5 The fragment of overlay visualization (constructed via VOSViewer v.1.6.15) [4]

The biggest environmental hazard relates to the inflow of chemical impurities contained in phosphogypsum in the form of water-soluble and volatile compounds. The content of radioactive, rare-earth elements and heavy metals is directly proportional to their content in phosphate raw materials (Table 8.1).

Magmatic phosphates (Colsky and South African apatite) have a higher content of rare earth elements (REEs) than sedimentary phosphates (Morocco, Florida, Senegal, etc.), and cadmium has a lower content of rare earth elements (REEs) [6–10].

Among the well-known areas of recycling and disposal of phosphogypsum, it is quite new to consider using phosphogypsum in environmental protection technologies.

Zhang et al. [11] conducted a study on developing an energy-efficient methodology for pure calcium carbonate production and the binding of CO_2 minerals by hydrolyzate of phosphogypsum decomposition products. Pure calcium carbonate was obtained by extracting and mineralising CO_2 using the hydrolyzate solution of phosphogypsum decomposition products. The carbonization rate of over 90% was obtained when the absorption of the liquid phase reached saturation. This method

Table 8.1 The Chemical composition of phosphogypsum converted to dihydrate depending on the type of raw material (in %)

Phosphogypsum sample	Origin of raw materials			
	Aktyubinsky (Tatarstan)	Apatite—phosphorite (South African)	Apatite (Colsky Peninsula)	Karatau (Kazakhstan)
CaO	24.4	30.0	32.4	31.2
SO_3	34.8	39.8	46.2	40.2
$P_2O_{5(total)}$	1.9	6.2	1.3	1.6
$P_2O_{5(water\ soluble)}$	1.1	3.3	0.7	0.8
Fe_2O_3	0.9	0.7	0.1	0.1
Al_2O_3	0.8	0.8	0.3	0.4
F	0.1	0.3	0.3	0.3
Insoluble residue	21.7	4.3	0.7	8.1
Crystalline water	15.6	17.4	18.6	18.0

can be referred to as an alternative solution to the environmentally friendly use of phosphogypsum.

A study by Bulat et al. [12] substantiates the expediency of creating X-ray protective structures from composite material based on phosphogypsum. However, the influence mechanisms of composite materials based on phosphogypsum binders require further studies on the efficiency of protection against X-ray and gamma radiation, which also requires further studies.

8.3 Involvement of Phosphogypsum in Bioprocesses to Produce Bio-based Products: A Development Perspective

The application of biotechnological methods in phosphogypsum utilization is crucial, as it supports the transformation of its components by various groups of microorganisms in environmental protection technologies. This approach provides a scientific basis for the effective conversion of phosphogypsum into valuable by-products, enhancing its role in sustainable waste management and pollution mitigation.

Bioremediation presents a viable strategy for managing phosphogypsum by leveraging living organisms, particularly microorganisms, to break down or neutralize environmental pollutants. This approach offers a cost-effective and environmentally safe alternative to conventional disposal methods. Various microbial groups can play a role in these bioprocesses. Trifi et al. [13] investigated the microbial communities present in a phosphogypsum dump in Tunisia. Their study identified autochthonous microorganisms within the accumulated phosphogypsum, including bacteria capable of recovering radioactive elements and heavy metals, as well as facilitating sulfate oxidation and reduction. These findings highlight the potential of microbial-based remediation in mitigating the environmental risks associated with phosphogypsum waste.

In the study by Barakhnina et al. [14], it was found that during three days of cultivation in the experiment with the addition of 1% wt. of phosphogypsum, the degree of destruction of carboxymethyl cellulose and polyacrylamide in drilling wastewater, was higher (99.8%) than in the control (66.8%). However, this method requires further development and improvement as the implementation on an industrial scale requires activation of the biological component of the process, considering all physical and chemical properties of biodegradable material. Another area of phosphogypsum use is its application in biotechnological processes of sewage sludge neutralization. The paper by Chernysh [15] describes a method of sewage sludge treatment in systems of anaerobic fermentation with heavy metals deposition into complex sulfide fraction with the help of biogenic hydrogen sulfide—a product of the life activity of sulfate reducers, where phosphogypsum is a mineral additive for intensification of bacterial culture development. To enhance the development of the bacterial culture, it is advisable to immobilize the bacteria on a mineral carrier. This approach reduces biomass loss from the bioreactor and enables a continuous treatment process, ultimately improving the bioreactor's efficiency and productivity.

Anaerobic fermentation combined with sulfate reduction presents a promising approach for phosphogypsum utilization. This method was explored in studies by Plyatsuk and Chernysh [16, 17], and Chernysh et al. [4] and further supported by Zouch et al. [18], who observed significant sulfide formation in enrichment cultures derived from marine sediments near the wastewater discharge point of a Tunisian fertilizer plant (Sfax, Tunisia). Species of Desulfovibrio have been identified as key candidates for phosphogypsum bioremediation, as they facilitate metal and metalloid immobilization through sulfide production.

Per previous studies, the biotechnologies of processing complex organic substances allow anaerobic as well as aerobic conditions to carry out the decomposition of organic wastes with the formation of volatile fatty acids and carbon dioxide, which in the liquid phase is represented by carbon dioxide [7, 19, 20]. Developing the biochemical process of converting phosphate raw materials into phosphate fertilizers is a promising field of study.

Combining recycling areas of already accumulated phosphogypsum and implementing new technological solutions for processing phosphorus raw materials will reduce the technogenic impact of phosphogypsum on the environment. Therefore, we have formed a model reflecting an integrated approach to environmentally friendly production and recycling of phosphogypsum (Fig. 8.6).

It is noteworthy that using a complex compost mixture of phosphogypsum, elementary superphosphate, manure, and plant residues in reclaimed soils can significantly reduce the levels of mobile forms of heavy metals (Co, Mn, Cu, Ni, Pb) by 60–70% or more within a year. This is achieved through the transformation of heavy metals into organic complexes and metal salts, which are less accessible to plants [21]. Phosphogypsum can be regarded as an acid-resistant mineral carrier and a

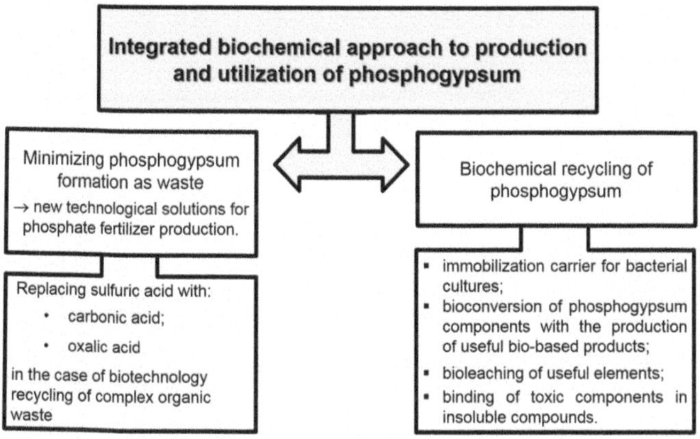

Fig. 8.6 An integrated biochemical approach to the production and utilization of phosphogypsum [4]

source of essential macro- and trace elements that support the growth of beneficial ecological and trophic bacterial groups. When used as an immobilizing carrier, phosphogypsum with a modified surface exhibits low porosity, while also promoting the formation of a stable biofilm on the surface of the granules. Thus, phosphogypsum holds promise as a carrier for bacterial immobilization in bioreactors.

8.3.1 Modeling and Analysis of the Effect of Phosphogypsum Features Under Estimation of E. coli Growth

Figure 8.7 shows the growth of *Escherichia coli* (*E. coli*) in LB broth with different doses of phosphogypsum. To obtain these data, the average values of the change in the OD_{600} value in a series of experiments were found. At the beginning of the experiment, when analyzing the same sample (measuring one sample three times at different doses of phosphogypsum), very different OD_{600} values were observed under the same conditions since the precipitation of the phosphogypsum components during mixing influenced the measurement readings. Within this experimental work, the primary OD_{600} of the culture medium with different phosphogypsum concentrations were considered. However, in subsequent measurements, there was still a significant variation in the readings of the OD_{600} value. The rapid sedimentation of particles of phosphogypsum components facilitated this. Therefore, only the supernatant was taken in further experiments. During the first 5 h of *E. coli* cultivation, an OD_{600} increase was observed in samples with phosphogypsum compared with the culture medium without its addition.

It is very important to develop a methodological approach for assessing the biochemical activity of phosphogypsum with an understanding of the chemical features of phosphogypsum under metabolic pathways and OD_{600} measurement.

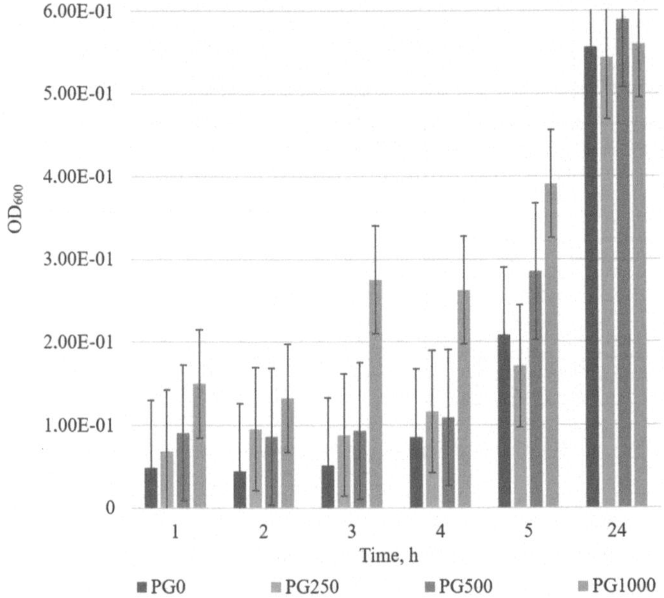

Fig. 8.7 OD_{600} cultural suspension at different doses of phosphogypsum [20]

Since phosphogypsum is a poorly soluble compound, its solubility in water does not exceed 1000 mg per liter. Such a low solubility in water is because phosphogypsum is formed from gypsum crystals, and it also has inclusions of aluminosilicates (Fig. 8.8). Ca, S, Si, Al, and P are the most represented elements in the phosphogypsum composition (Fig. 8.9).

However, the metabolic activity of bacteria can affect the solubility of phosphogypsum. Several factors influencing bacterial growth under this system are discussed: (i) the initial increase in the turbidity of the culture medium with different doses of phosphogypsum and the rapid deposition of phosphogypsum components on bottom of the tube with decrease in the turbidity; (ii) partial splitting of the components of phosphogypsum in the medium during the time of cultivation of *E. coli* and reducing sediment; (iii) after 24 h, tubes with culture media with phosphogypsum and without

Fig. 8.8 SEM images of phosphogypsum sample [20]

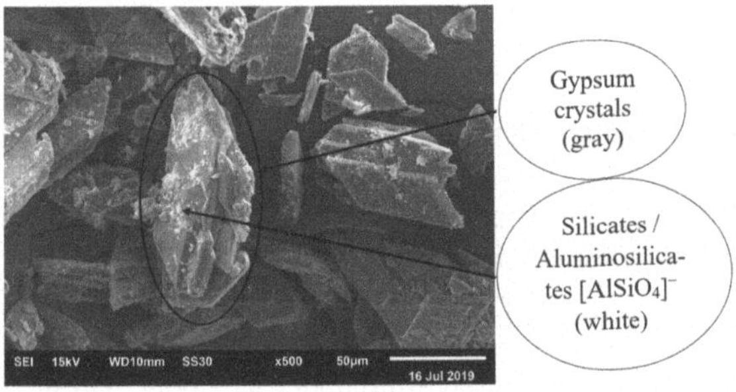

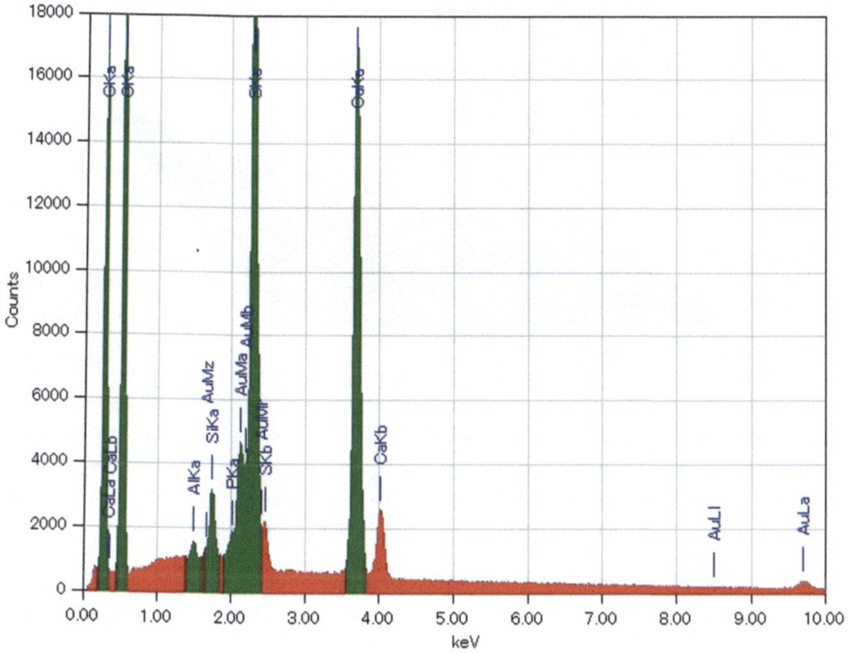

Fig. 8.9 EDS spectra of the field (Au is presented only under the metallization process of the sample). The horizontal scale represents X-ray energy (keV), green elemental peaks indicate the highest intensities, and red elemental peaks are closer to the background level [20]

it had approximately the same turbidity (see Fig. 8.10) and this was confirmed when evaluating the optical density (see Fig. 8.7).

Thus, the important task is the correct assessment of the effect of phosphogypsum properties on a possible increase in the turbidity of the nutrient medium of some

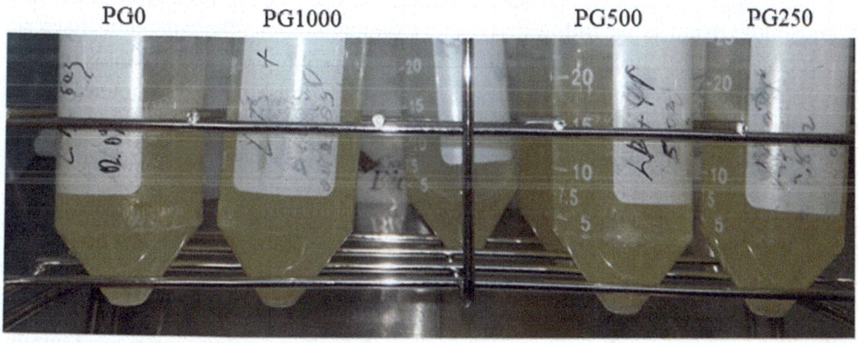

Fig. 8.10 Photography tubes after 24 h of *E. coli* cultivation [20]

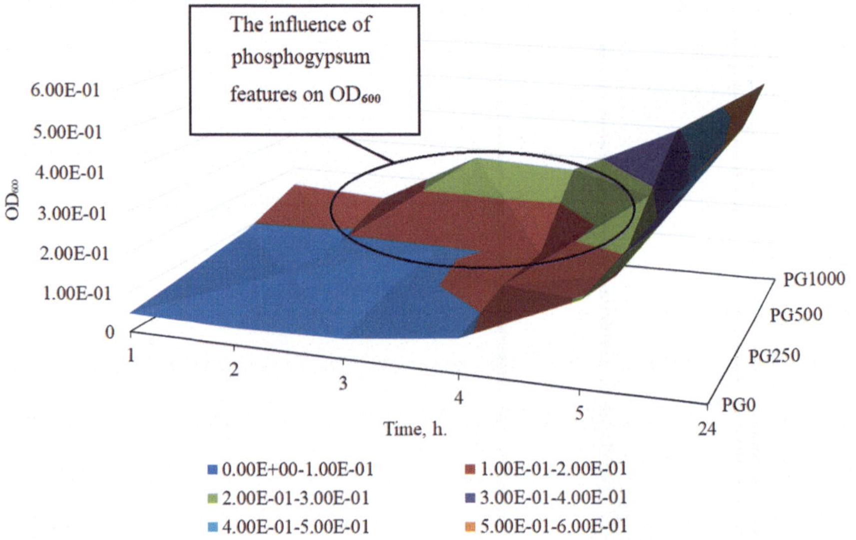

Fig. 8.11 The 3D surface under 'culture-medium-phosphogypsum' system [20]

samples when raising this precipitate with stirring, which affects the evaluation of the experimental results.

Such effects of the influence of phosphogypsum can be seen on the three-dimensional surface of the change in the OD_{600} of the culture suspension (Fig. 8.11), which indicates the need for a series of experiments considering the above-mentioned observed properties in the 'culture-medium-phosphogypsum' system. This will allow the most objective assessment of the phosphogypsum effect on *E. coli* growth.

Figure 8.12 shows the general methodological approach for assessing the biochemical activity of phosphogypsum. It is also important to evaluate the change in the solubility of phosphogypsum components upon exposure to metabolites.

The methodological idea of model systems to develop eco-friendly bio-utilization of phosphogypsum is as follows: (i) select biotechnologically amenable bacteria as models; (ii) check bacteria responses against phosphogypsum; (iii) analyse their responses genetically and theoretically; (iv) apply these findings to the bacteria which are used in the wastewater treatment plant.

The research by Chernysh and Hasegawa [20] uses *E. coli* as a model to estimate its response against phosphogypsum effluence. It should be noted that a separate study of the metabolic pathways of strict anaerobes is necessary. For example, sulfate-reducing bacteria (SRB) use sulfate as an electron acceptor and SRB can be involved in a consortium with methanogenic bacteria during anaerobic digestion.

Biofilms, in this case, are preferable to planktonic cultures because, as mentioned earlier, these types of cultures are more resistant to changing environmental conditions. Microorganisms in biofilms, protected by a layer of extracellular matrix, are better adapted and much more resistant to high concentrations of various groups of xenobiotics, particularly chlorine and nitrogen-containing aromatic compounds. In

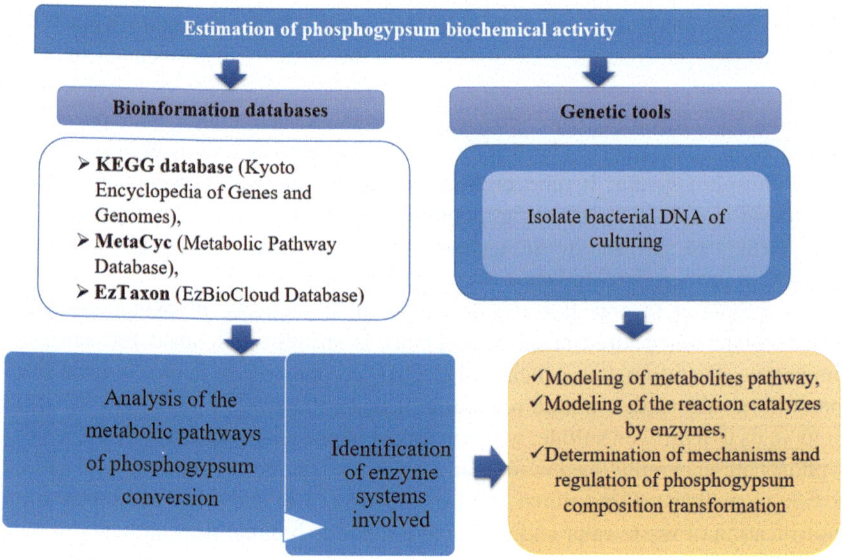

Fig. 8.12 A general model of estimation bacteria growth under a medium that included Phosphogypsum [20]

some cases, the matrix is involved in bioremediation, sorbing and retaining toxic substances from the aqueous phase.

Thus, the development of an approach to the treatment of wastewater and chemical wastes such as phosphogypsum for coagulation and aggregation to achieve the effective removal of biogenic and toxic substances using multidimensional biofilms containing, in addition to hydrocarbon-oxidizing bacteria, other ecologically trophic groups, for example, *E. coli*, is one of the possible areas for further research.

8.3.2 Biotreatment of Sewage Sludge with Phosphogypsum Under Anaerobic Condition: Phosphorus Removal

Phosphorus is a crucial nutrient with the smallest reserves and cannot be replaced by other elements. Based on research findings, scientific principles regarding the mechanisms of phosphorus removal from sludge and sewage sludge during anaerobic stabilization under sulfate reduction, in conjunction with phosphogypsum, have been developed. A key aspect of this process involves not only the release of phosphate ions from the microbial community of active sludge during anaerobic fermentation, but also the breakdown of metal-phosphate complexes in sewage sludge (SS). It is important to note that many heterotrophic microorganisms in wastewater and active sludge have the ability to accumulate phosphorus. During anaerobic stabilization, phosphate ions diffuse from microbial cells as volatile fatty acids are actively released. Additionally, the lysis of these microbial cells during anaerobic degradation can release polyphosphates, as demonstrated in the study by Plyatsuk and Chernysh [15].

Figure 8.13 illustrates the substance flow model associated with phosphate removal from wastewater sludge during combined anaerobic stabilization with dehydrated phosphogypsum. It reflects the dynamics of the sulfate-reducing community's development in the bioreactor, considering the process of returning a portion of the fermented substrate to the technological system, as well as the behavior of flocculation processes within the sewage sludge biomass.

It is important to note that the thermodynamic sequence of oxidation–reduction potentials for desired electron acceptors is strictly maintained for inorganic compounds. Therefore, the sequence of electron acceptor usage in biochemical reactions follows this order: nitrate reduction or denitrification → reduction of sulfur compounds to hydrogen sulfide, which is linked to phosphorus release from the cells of phosphate-accumulating microorganisms in the sludge → reduction of carbon dioxide to methane. For optimal sulfate reduction, it is necessary to first perform denitrification of wastewater and sludge in a separate biotreatment unit.

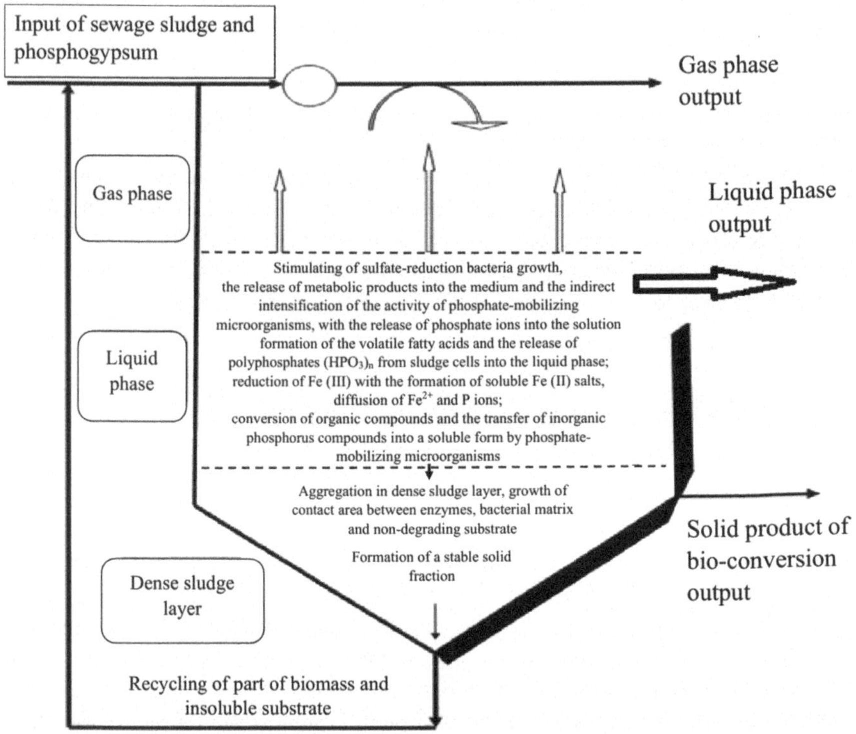

Fig. 8.13 Block diagram of spatial distribution in the bioreactor during dephosphatation of sewage sludge together with phosphogypsum under condition of dissimilation sulfate reduction [22]

In general, the patterns of biochemical transformations of phosphorus in the anaerobic sewage sludge and phosphogypsum conversion are as follows:

1. Mobilization of phosphorus from its mineral compounds, in particular, Fe (III) hydroxide, which bound P, appears in the environment deprived of oxygen supply (anaerobic bioreactor), and Fe (III) is reduced by organic matter found in sewage sludge in the process of microbial reduction. Since Fe (II) salts, including $Fe_3(PO_4)_2$, are soluble, then ions of Fe $^{2+}$ and P begin to diffuse.
2. Under anaerobic conditions, due to the fermentation of organic substances, the formation of volatile fatty acids and release of polyphosphates $(HPO_3)_n$ from the sludge cells into the liquid phase occur.

It should be noted that phosphate ions, having passed into the solution, react with calcium ions, forming low-soluble compounds.

The mechanisms of anaerobic stabilization under sulfate reduction conditions with phosphorus recovery can be described with the following biochemical reactions of phosphate removal:

$$2CH_3COOH + (HPO_3)_n + H_2O \rightarrow (C_2H_4O_2)_2 + a_{PO}PO_4^{3-} + a_H H^+ \quad (8.1)$$

$$Fe_3(PO_4)_2 + 3SO_4^{2-} + 6C_{org} \rightarrow 3FeS \downarrow + 2PO_4^{3-} + 6CO_2, \quad (8.2)$$

where C_{org}—organic substrate; a_{PO}, a_H—stoichiometric coefficients.

The biochemical equations describe the terminal stage of sewage sludge organic matter conversion. When introducing macro- and microelements (phosphogypsum) into the source system, the process of sulfate reduction is intensified, and consequently, the development of ecological and trophic groups of bacteria performing it corresponds to the auto-selection principle. Herewith, during the sulphate reduction, polyphosphates $(HPO_3)_n$ are released from sludge cells into the liquid phase, which is more simply described in Eq. (8.1) where $(C_2H_4O_2)_2$ is a reserve organic matter that is present in sewage sludge liquid phase.

Equation (8.2) describes the reduction reaction of Fe (III) phosphate under anaerobic conditions to ferrous iron with the subsequent binding of Fe^{2+} ions in the sulphide form. It should be noted that the ions of metals actively interact with hydrosulfide and sulfide ions. In contrast, phosphate ions are emitted into the liquid phase, from which phosphorus can be recovered in the form of its salts with the subsequent use as a fertilizer.

During anaerobic stabilization under the influence of sulfate-reducing agents, organic matter (carbohydrates, organic acids) undergoes dehydrogenation; hydrogen is transferred to sulfates, sulfites, or thiosulfates, which are reduced to hydrogen sulfide. Hydrogen sulfide actively interacts with the solid phase of the sludge and with cations in the liquid phase. Throughout this anaerobic process, the organic component of the sewage sludge undergoes mineralization, and chelate complexes with heavy metals are transformed into soluble organic compounds. This results in increased mobility of heavy metals, which migrate into the liquid phase of the sludge, ultimately precipitating as an insoluble fraction, significantly altering the sludge's properties. Additionally, heavy metals that form salts with the organic components of sewage sludge are exchanged for calcium ions via ion exchange on the surface of phosphogypsum. As a result, during anaerobic stabilization, organic chelate complexes with heavy metals are broken down, and isomorphic compounds such as metal sulfides (e.g., iron sulfide—marcasite, zinc sulfide—sphalerite, nickel sulfide) and carbonates are formed. A comparative elemental and phase analysis of the final treatment product is presented in Table 8.2.

Precipitation of heavy metals in the form of a complex sulphide fraction is an important factor for the stable functioning of sulfate reducer association in the bioreactor.

The analysis of the biochemical processes proceeding during the anaerobic stabilization of the sewage sludge/wastewater with dehydrated phosphogypsum is incorrect from the point of view of their simple summation. They define a new level of self-organization of the system, receiving new qualitative characteristics that determine the environmental safety of the sewage sludge and wastewater processing product. This represents the synergistic effect of '2 + 2 = 5', meaning that the outcome exceeds the simple sum of its parts. Every biochemical reaction proceeds in coordination with various factors and other reactions (chemical, physical, biological) in the system that determine each other. Consequently, the optimal variant of system development based on self-organization principles is formed, which is appropriate to natural environmental systems.

The process of anaerobic stabilization forms an organo-mineral product which, after the separation from the liquid fraction and drying, has a grey-brown colour, and its consistency reminds of aggregated soil particles (Fig. 8.14).

Table 8.2 Results of sewage sludge samples analysis of solid phase together with phosphogypsum after anaerobic stabilization

Element	Phase composition
Si P S Ca K Mg Al Fe Ni Cu Zn Cr	JCPDS 77-1060 (quartz)
	JCPDS 74-1433 (gypsum)
	JCPDS 26-905 (potassium hydrogen phosphate hydrate)
	JCPDS 11-293 (brushite)
	JCPDS 40-660 (ammonium sulfate)
	JCPDS 86-2342 (calcium carbonate)
	JCPDS 79-43 (zinc sulfide)
	JCPDS 37-475, 74-1051 (sulfides of iron)
	JCPDS 26-1116, 65-3928 (sulfides of cuprum)
	JCPDS 65-395 (nickel sulfide)
	JCPDS 10-345 (chromium sulphide)
	JCPDS 89-1460 (calcium aluminum silicate)

Fig. 8.14 The general view of the organo-mineral composite material based on phosphogypsum and sewage sludge after the anaerobic stabilization under conditions of dissimilating sulfate reduction [23]

Such components as quartz or amorphous silica—SiO_2; potassium hydroxide KOH; brushite—$CaPO_3(OH)\cdot 2H_2O$; calcide—$CaCO_3$; ammonium sulfate (mascagnite); iron sulfide (marcasite), zinc sulfide (sphalerite), copper sulfide (indigo copper), chromium sulfide, etc., forming a complex sulphide group (Table 8.2) are among the main components of the mineral constituents of the end product of the mixture anaerobic stabilization. The content of the main elements in the organo-mineral composite based on phosphogypsum and sewage sludge obtained after their joint anaerobic stabilization under dissimilating sulfate reduction is presented in the Table 8.3.

The phosphorus compounds have been released in a liquid fraction and can be applied with the obtained organo-mineral product. The results of the analysis of the mineral component of the end product confirm the main statements of the above mentioned biochemical model of the process.

Table 8.3 Content of the main elements in biogenic composites expressed as oxides, % [22]

Component	Content (wt%)
SiO_2	37.65
Al_2O_3	8.4
CaO	31.3
K_2O	3.2
MgO	1.7
Na_2O	3.4
SO_3	9.5
ZnO	0.01
Fe_2O_3	0.45
CuO	0.003
P_2O_5	0.5
NiO	0.0014

Phosphogypsum has a beneficial effect on enhancing the quality of digestate and can be utilized as an organic mineral fertilizer enriched with mineral components [24]. While it is clear that adding phosphogypsum to plant waste does not significantly improve biogas production, there is potential for analyzing its use as an additive with animal waste, which has a higher content of nitrogen compounds.

8.3.3 Biotechnological Solutions of Phosphogypsum Recycling in the Wastewater Treatment Plant

The sequence of thermodynamic oxidation–reduction potentials as preferred acceptors is applied to inorganic compounds. The sequence of electron acceptor use includes:

1. Nitrate reduction or denitrification.
2. Reduction of sulfur compounds or sulfate reduction associated with phosphorus release from the cells of phosphorus-accumulating sludge microorganisms.
3. Reduction of carbon dioxide to methane or methanogenesis.
4. Thus, wastewater and sewage sludge should be denitrified in a separate biological treatment unit for the sulfate reduction process to progress better. Important provisions required to build a process flow diagram are:
    1. Predenitrification of sludge deposits with a portion of a wastewater stream improves anaerobiosis in the bioreactor (tank).
    2. Phosphorus removal is intensified with the introduction of phosphogypsum into the anaerobic tank, which contributes to the beginning of the sulfate reduction process on the principles of auto selection.
    3. Removing phosphate ions from the liquid phase in a separate treatment unit using chemical deposition is more reasonable.

Figure 8.15 shows the process flow diagram.

Certain technology provisions are patented inventions; a utility patent covers the design of the anaerobic unit for "Unit for anaerobic treatment of organic waste".

Denitrification, the first step of the process, occurs in a biological unit that receives sludge deposits along with a portion of the wastewater flow. The second stage involves anaerobic microbial degradation under sulfate reduction, which takes place after transferring the sludge deposits from the biological unit into the anaerobic tank, where phosphogypsum is introduced, and gas is extracted from the system for later use. The third stage is phase separation during the settling process. In this phase, the solid phase is removed from the system, while the liquid phase is directed to the phosphorus removal unit (fourth stage), where dissolved phosphorus compounds are precipitated using an agent like calcium hydroxide. Phosphate ions can be precipitated from the liquid phase of sludge deposits with various reagents and subsequently used as agricultural fertilizers. It is important to note that the phosphate sorption mechanism in sludge deposits is enhanced by the binding of phosphate ions

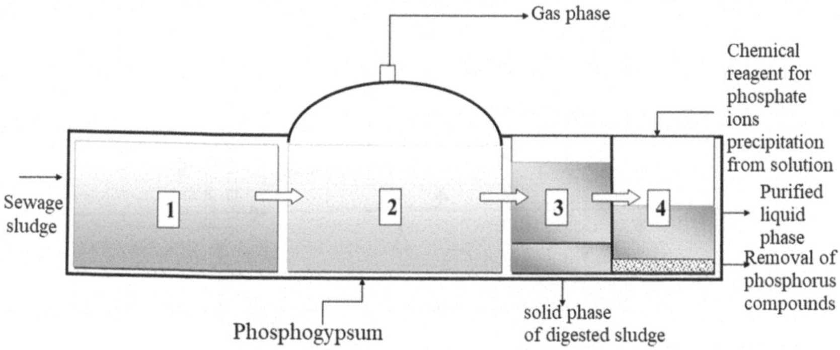

Fig. 8.15 Schematic of biological treatment of sewage sludge together with phosphogypsum [23]. 1—denitrification unit, 2—anaerobic reactor of dephosphatation treatment under the sulfate reduction conditions; 3—sedimentation tank; 4—reagent phosphorus removal unit

with metal hydroxides. Additionally, phosphorus is incorporated into the hydrated potassium complexes found in sludge from sludge maps.

The startup of the wastewater and sewage sludge treatment unit requires the presence of an initial amount of destructor microorganisms, which act as an inoculum. The main stages of the startup process include: cultivating the essential ecological and trophic groups using fermenters under laboratory conditions; separating viable biomass; introducing and immobilizing the necessary microbial groups at anaerobic (B2 and B3) and aerobic (B1) stages, which may be located in a separate aeration chamber where activated sludge from existing urban treatment plants can be inoculated; filling with wastewater and sewage sludge, promoting the development of a microbial community formed through auto-selection by adding dihydrate phosphogypsum; and transitioning the unit to a flow-through mode.

Figure 8.16 presents a process flow diagram of wastewater and sewage sludge treatment units with phosphogypsum recycling.

The developed process flow diagram includes a three-stage treatment process with a high-rate first stage of aerobic oxidation with a sludge age of 0.5 days and hydraulic retention time of 0.45 h. The efficiency of COD removal can reach 32–50% at this stage. With a short retention time of 0.45 h, organic matter is only sorbed to active sludge and does not have enough time to be absorbed by the cell, which has a positive impact on further gas generation in anaerobic bioreactors (B_2 and B_3).

The denitrification stage occurs in B_2 and is carried out with a maintained activated sludge age of 5–10 days, depending on the temperature. The dissolved oxygen level is maintained within the 0.1–0.97 mg/dm^3 range. Hydraulic retention time runs at 1–2 days under psychrophilic conditions (20 °C) and 0.5–1 days under mesophilic conditions (35 °C). The longer the hydraulic retention time of the substrate (fresh sludge), the higher the potential COD output of the dissolved substance. It has been determined that under a stationary mode, the output of dissolved organic matter from B_2 with an extended sludge bed (hydrolysate and the residue of dissolved urban wastewater matter) can reach up to 30% of the biodegradable wastewater substance

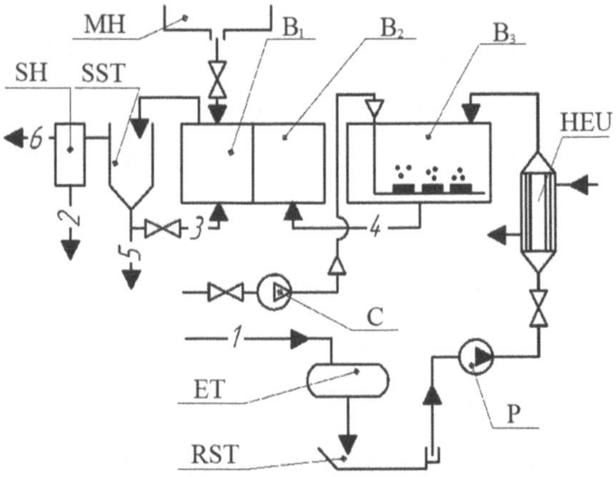

Fig. 8.16 Process flow diagram of wastewater and sewage sludge treatment using phosphogypsum. 1—input stream to be treated; 2—sludge containing precipitated phosphorus compounds; 3—recycling of a portion of digested sewage sludge; 4—nitrate reverse cycle; 5—digested sewage sludge to be recycled; 6—treated liquid phase; ET—equalization tank; RST—radial-flow settling tank; P—pump; HEU—heat exchange unit; B_1—aerobic bioreactor; B_2—anaerobic bioreactor for the denitrification stage; B_3—anaerobic bioreactor for the phosphorus removal stage under sulfate reduction with the release of phosphorus compounds; MH—measuring hopper for phosphogypsum; SST—secondary settling tank; SH—settling hopper for reactant removal of phosphorus from the liquid phase; C—compressor [25]

COD within half a day. Sewage sludge from previous stages of biological treatment is digested during the phosphorus removal stage under sulfate reduction, which occurs in B_3. The formation rate of volatile fatty acids and acetates and phosphorus removal efficiency reaches maximum values when the optimum sludge age is between 5 and 11 days. Thus, under mesophilic conditions, sludge retention time (including the initial induction period of not less than 5 days, which is only observed upon unit startup) is required to achieve the indicators of organic matter decomposition up to 65%, and the degree of phosphate anion release not less than 70%, equals 10 days. The specific level of gas generation equals over 650 dm^3/g of the benzene substance. The hydraulic retention time of wastewater equals 1.5–2 h.

The immobilization efficiency criterion of microbial cells based on the repeated increase of the biomass yield of free cells indicates the saturation of the phosphogypsum medium with a biomass of destructor microorganisms. It should be noted that most of the microorganisms are immobilized within the first 5 days, and then, with an increase in contact time, the degree of cell adsorption decreases dramatically. At that, phosphogypsum recycling occurs of about 6.5 kg/m^3 of sewage sludge sent for treatment.

The wastewater and sewage sludge treatment technology that includes an aerobic-anaerobic conversion process provides a spatial microbial succession and a trophic hydrobiont levels involving technogenic mineral resources (e.g. phosphogypsum) to stimulate the development of essential ecological and trophic groups.

The following parameters are monitored to control the efficiency of the treatment process: wastewater temperature upon anaerobic bioreactor entry; pH value of the environment upon entering and leaving anaerobic and aerobic bioreactors; dissolved oxygen concentrations (pO_2) upon entering and leaving the aerobic bioreactor; oxidation–reduction potential value (Eh) in the anaerobic bioreactor; station pumping unit alarm; tank fluid level alarm; phosphate concentration (PO_4^{3-}) upon entering and leaving anaerobic and aerobic bioreactors. After phosphate precipitation, the treated water in the settling hopper can accumulate in accumulation tank sections and run by gravity into the pump area for treated wastewater from urban treatment plants [22].

In order to improve microbial living conditions and, correspondingly, the effect of wastewater and sewage sludge treatment, the anaerobic bioreactor is introduced with air coming from air blowers, thus saturating (aerating) wastewater with atmospheric oxygen. Air is supplied into the airlift area of each section through perforated pipelines. During the aerobic stage, oxygenated wastewater circulates through the operating area B_1 and is purified by microbes immobilized on fibrous media. B_1 is aerated by compressor-induced air from the atmosphere and gases released from B_2.

Battery-type anaerobic bioreactors (B_2 and B_3). These consist of sections (at least 23 for each one) interconnected in series with a gravity flow mode. A bioreactor section is a vertical plastic pipe with a diameter of 350 mm and a height of 4.5 m. Each B_3 section is automatically supplied with phosphogypsum from the measuring hopper. Gas outlet manifolds connected with reactor sections are mounted in the upper part of units. The area between bioreactor walls and protective polystyrene sheet shields serves as an airlift area. Upon its circulation, activated sludge in B_3 coagulates with phosphogypsum particles, which is accompanied by matrix forming and allows the reduction of the loss of active biomass from the system. At that, low-grade wastewater heat may be used.

The methodology based on equalising wastewater COD mass balance following Chernysh et al. [4] is commonly used for designing biochemical wastewater treatment processes. This methodology, however, does not consider the physical and chemical properties of mineral media that can be used for wastewater and sewage sludge treatment and the indicators of joint development of ecological and trophic microbial groups in one community. Accordingly, these factors were taken into account, and the approach for calculating the volume of the anaerobic bioreactor for the phosphorus removal stage in m^3 was created:

$$V_{AnB} = \left(\frac{w_{SS} \cdot m_{SS}}{\rho_{SS} \cdot 100\%} + \frac{m_{PG}}{\rho_{PG}}\right) \cdot \left(\frac{Y_{X/(S+D)} \cdot (\Delta C_{COD} + \Delta C_P)}{\tau(1 - Y_{X/(S+D)}) \cdot \sum \mu_{iMC} \cdot X_{gen}}\right) \quad (8.3)$$

where m_{SS} is mass of sewage sludge that enters the phosphorus removal stage, kg; w_{SS} is sewage sludge moisture, %; m_{PG} is phosphogypsum mass, kg; ρ_{SS} and ρ_{PG} are respective sewage sludge and phosphogypsum density, kg/m^3; ΔC_{COD} is COD load, kg/m^3; ΔC_P is phosphate load, kg/m^3; τ is time spent at the phosphorus removal stage, days; μ_{iMC} is specific growth rate of *i*-type of microbial communities, $days^{-1}$; X_{gen} is total productive biomass in the bioreactor, kg/m^3; $Y_{X/(S+D)}$ is economic coefficient of biomass yield based on a complex substrate (COD/Phosphogypsum).

Indicators of treatment efficiency in the B_1–B_2–B_3 system can reach 95.5% for COD and 99.5% for phosphates. At that, not less than 60% of COD and 70% of phosphates are destroyed during anaerobic stages (denitrification and phosphate removal under sulfate reduction).

It should be noted that a typical pattern lies in the fact that when nitrates predominate the content of wastewater and sewage sludge, it is advisable to initially provide the anaerobic stages (B_2–B_3), where the organic substrate is oxidized at the expense of nitrates with the release of free nitrogen. In contrast, the aeration stage should be provided for final oxidation of organic substances.

References

1. El Kateb A, Stalder C, Rüggeberg A, Neururer C, Spangenberg JE, Spezzaferri S (2018) Impact of industrial phosphate waste discharge on the marine environment in the Gulf of Gabes (Tunisia). PLoS ONE 13:e0197731
2. Gaudry A, Zeroual S, Gaie-Levrel F, Moskura M, Boujrhal FZ, El Moursil RC et al (2007) Heavy metals pollution of the Atlantic marine environment by the Moroccan phosphate industry. Water Air Soil Pollut 178:267–285
3. El Zrelli R, Rabaoui L, Abda H et al (2019) Characterization of the role of phosphogypsum foam in the transport of metals and radionuclides in the Southern Mediterranean Sea. J Hazard Mater 363:258–267. https://doi.org/10.1016/j.jhazmat.2018.09.083
4. Chernysh Y, Yakhnenko O, Chubur V, Roubík H (2021) Phosphogypsum recycling: a review of environmental issues, current trends, and prospects. Appl Sci 11(4):1575. https://doi.org/10.3390/app11041575
5. Haneklaus N, Barbossa S, Basallote MD, Bertau M, Bilal E, Chajduk E, Chernysh Y, Chubur V et al (2022) Closing the upcoming EU gypsum gap with phosphogypsum. Resour Conserv Recycl 182:106328. https://doi.org/10.1016/j.resconrec.2022.106328
6. Nadymova MA (2016) Apatite concentrate as a promising source of rare-earth metals in complex processing. In: Proceedings of the international scientific and practical conference "The North and the Arctic in the new global development paradigm. Luzin Readings". Luzin Institute for Economic Studies, pp 674–679
7. Chernysh Y, Plyatsuk L (2020) Environmentally friendly concept of phosphogypsum recycling on the basis of the biotechnological approach. In: Leal Filho W, Borges de Brito P, Frankenberger F (eds) International business, trade and institutional sustainability. Springer, Cham, pp 167–182
8. Myrzakhmetov B, Sholak A (2015) Chemical composition and complex processing method of phosphogypsum. In: Proceedings of the West Siberian scientific center conference: modern trends in the development of science and production. Kemerovo, pp 67–71
9. Vorobiev AE, Danilova EV (2013) Audit of rare-earth apatite ore processing technology. RUDN J Eng Res 4:36–40
10. Samonov AE, Vanshin YV (2007) Problems with apatite processing. Geol Geogr Global Energy Sci Technol J 2:7–9
11. Zhang W, Zhang F, Ma L, Ning P, Yang J, Wei Y (2020) An efficient methodology to use hydrolysate of phosphogypsum decomposition products for CO_2 mineral sequestration and calcium carbonate production. J Clean Prod 259:120826. https://doi.org/10.1016/j.jclepro.2020.120826
12. Bulat AF, Ivanov VA, Holov KS, Mysovets YV (2010) Radio-protective properties of phosphogypsum binding agent with rare-earth filler. Sci Bull Natl Mining Univ 5:48–51

13. Trifi H, Najjari A, Achouak W, Barakat M, Ghedira K, Mrad F et al (2019) Metataxonomics of Tunisian phosphogypsum based on five bioinformatics pipelines: insights for bioremediation. Genomics 112(1):981–989. https://doi.org/10.1016/j.ygeno.2019.06.014
14. Barakhnina VB, Khafizova AA, Kireev IR (2011) The investigation of using phosphogypsum in drilling discharged waters biopurification. Bashkir Chem J 18(2):90–92
15. Chernysh YY (2014) Sewage sludge utilization by sulfidogenic association of microorganisms (Doctoral dissertation). Sumy State University, Sumy
16. Plyatsuk LD, Chernysh YY (2014) Intensification of the anaerobic microbiological degradation of sewage sludge under bio-sulfidogenic conditions. J Solid Waste Technol Manage 40(1):20–23
17. Plyatsuk LD, Chernysh YY (2014) Intensification of the anaerobic microbiological degradation of sewage sludge under bio-sulfidogenic conditions. J Solid Waste Technol Manage 40:10–23
18. Zouch H, Karray F, Armougom F, Chifflet S, Hirschler-Réa A, Kharrat H et al (2017) Microbial diversity in sulfate-reducing marine sediment enrichment cultures associated with anaerobic biotransformation of coastal stockpiled phosphogypsum (Sfax, Tunisia). Front Microbiol 8:1583. https://doi.org/10.3389/fmicb.2017.01583
19. Plyatsuk LD, Chernysh YY (2016) The removal of hydrogen sulfide in the biodesulfurization system using granulated phosphogypsum. Eurasian Chem Technol J 18(1):47–54. https://doi.org/10.18321/ectj395
20. Chernysh Y, Hasegawa K (2020) Improvement of the model system to develop eco-friendly bio-utilization of phosphogypsum. In: Lecture notes in mechanical engineering: advances in design, simulation and manufacturing III. Springer, pp 357–366
21. Denzanov GA, Petruk GD (2006) Resource-saving technology of bioconversion of natural phosphates. Environ Bull 2:25
22. Chernysh Y, Balintova M, Plyatsuk L, Holub M, Demcak S (2018) The influence of phosphogypsum addition on phosphorus release in biochemical treatment of sewage sludge. Int J Environ Res Public Health 15(6):1269. https://doi.org/10.3390/ijerph15061269
23. Chernysh YY (2021) Environmental biotechnologies for phosphogypsum recycling: protection of the atmosphere, hydrosphere and lithosphere. Sumy State University, Sumy
24. Chubur V, Chernysh Y, Ferchau E, Zaffar N (2022) Effect of phosphogypsum addition on methane yield in biogas and digestate properties during anaerobic digestion. J Eng Sci 9(1):H11–H18. https://doi.org/10.21272/jes.2022.9(1).h2
25. Chernysh EY (2019) Scientific principles of the ecological-synergetic approach to phosphogypsum utilization process for reducing technogenic pressure on the environment. Doctoral dissertation, Sumy State University, Sumy
26. Chernysh Y, Plyatsuk L, Dorda V (2014) Ecotechnology for hydrogen sulfide removal and production of elemental sulfur. Int J Energy Clean Environ 15(2–4):189–202

Conclusions

1. The principle of environmental biotechnology is based on addressing environmental problems by developing and applying biotechnological methods for waste recycling, water purification, and pollution prevention. The range of tasks addressed by environmental biotechnology is extremely broad—from the development and improvement of the methodology for comprehensive chemical and biological research of ecosystems near sources of anthropogenic impacts to the development of technologies and recommendations for soil reclamation, biological water and air purification, and the biosynthesis of products that mitigate the harmful effects of environmental changes on humans and animals.
2. Environmental biotechnology is a low-waste technology with environmentally friendly technical process implementation. They aim to clean the environment from various pollutants and produce environmentally friendly products with the possibility of recycling secondary resource flows. Therefore, the choice of biological agent has a conceptual basis associated with the need for in-depth knowledge of physiology and biochemical processes of growth and metabolism of biological agents and the need for their sanitary and hygienic assessment.
3. Substrates and media used in environmental biotechnology are very different; their range constantly expands. With the development of industrial processes, there is an accumulation of new types of waste that can be converted into useful products by applying biotechnology methods. On the one hand, traditional biotechnological industries, which are developing rapidly, may face the problem of depletion of traditional raw materials, so there is a need to expand the raw material base; on the other—increasing the amount of accumulated waste requires the development, of non-traditional methods (including biotechnological methods) for their processing and management.

Y. Chernysh et al., *Sustainable Biotechnology and Environmental Protection*, Green Energy and Technology, https://doi.org/10.1007/978-3-031-89436-7

4. Regarding the stages of biotechnological processes and their operational design, it should be noted that the maintenance and preparation of pure culture are critical for the pre-fermentation stage, as the biological agents determine the effectiveness of the entire biotechnological process through their physiological and biochemical characteristics and properties.
5. The fermentation stage in the biotechnological process utilizes the most complex and specific equipment that requires engineering calculations of the main biotechnological indicators. The most technically complex processes for fermentation are aerobic, deep sterile, and continuous (or with substrate feeding). The apparatus for surface and anaerobic fermentation is less complex and energy-intensive. The number of cultivation methods and the extraordinary variety of biological agents used have led to a huge variety of design solutions, depending on a number of factors: the type of biological agent and media, technology and scale of production, as well as the target product and others.
6. The range of products obtained in environmental biotechnological processes is extremely wide. Therefore, it is essential to consider the efficiency criteria of biotechnological product production. In terms of variety and volume of production, bioproducts obtained in processes based on the activity of microorganisms are in the first place.
7. The most important area of environmental safety is microbiological safety. In this direction, it is important to develop special engineering and technical support for biotechnological industries and the possibility of disposing of their waste. Wastes of biotechnological productions belong, as a rule, to the type decomposing in natural conditions under the influence of various factors (biological—mineralization with the participation of microorganisms; chemical—oxidation, physicochemical due to complex influence, for example, radiant energy and chemicals).

The manufacturer's authorised representative in the EU is Springer Nature Customer Service Centre GmbH, Europaplatz 3, 69115 Heidelberg, Germany. If you have any concerns regarding our products, please contact ProductSafety@springernature.com

Printed and bound by CPI Group (UK) Ltd, Croydon, CR0 4YY
01/10/2026
02225960-0002